Leonard Howard Lloyd Irby

The ornithology of the Straits of Gibraltar

Leonard Howard Lloyd Irby

The ornithology of the Straits of Gibraltar

ISBN/EAN: 9783743365162

Manufactured in Europe, USA, Canada, Australia, Japa

Cover: Foto ©berggeist007 / pixelio.de

Manufactured and distributed by brebook publishing software (www.brebook.com)

Leonard Howard Lloyd Irby

The ornithology of the Straits of Gibraltar

THE ORNITHOLOGY

OF THE

STRAITS OF GIBRALTAR.

BY

LIEUT.-COLONEL L. HOWARD L. IRBY, F.Z.S.,

H.-P. LATE SEVENTY-FOURTH HIGHLANDERS,
MEMBER OF THE BRITISH ORNITHOLOGISTS' UNION.

"Flumina amo sylvasque inglorius."

LONDON:

PUBLISHED BY R. H. PORTER,
6 TENTERDEN STREET, HANOVER SQUARE.

1875.

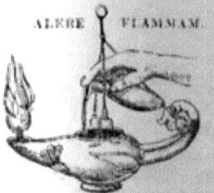

PRINTED BY TAYLOR AND FRANCIS,
RED LION COURT, FLEET STREET.

ERRATA ET CORRIGENDA.

Page line
20, 25, for *A. ferus* read *A. cinereus*
84, 17, for Gm. read Linn.
94, 30, for Penn. read Bodd.
96, 36, for upper read lower
96, 37, for lower read upper
106, 33, for Strike read Shrike
132, 34, for Woodpidgeon read Woodpigeon
196, 7, for *Anser ferus* read *Anser cinereus.*

ORNITHOLOGY

OF THE

STRAITS OF GIBRALTAR.

INTRODUCTION.

The list of birds and ornithological notes contained in this book are compiled from observations made on both sides of the Straits of Gibraltar—on the African side within a region extending from Tangier southwards to the lakes of Ras Doura beyond Larache, and eastward from Tangier to Tetuan and Ceuta, not reaching inland more than ten or twelve miles; on the European side I include that part of Andalucia which would be bounded by an imaginary line drawn from Gibraltar to Malaga, thence to Granada, Cordova, and Seville, along the delta of the Guadalquivir to Cadiz.

Nearly all the information relating to the birds of the Spanish side of the Straits is collected from personal observations made during a more or less prolonged stay at the Rock, between February 1868 and May 1872, and again from February to May 1874, but including during this time only one summer period, viz. July, August, and the first half of September. For the first three years of my residence at Gibraltar I was quartered with my regiment, the remaining time being passed there chiefly with a view to ornithological pursuits, from time to time making excursions, generally of about a fortnight's duration, to some part or other within the districts above mentioned, but chiefly confining my attentions to the country within a day's journey of Gibraltar.

The observations on the Moorish birds are in a great measure culled from the MS. of the late M. F. Favier, a French collector well known to the ornithological world, who, after a residence of about thirty-one years at Tangier, died there in 1867. I was informed that he had left a MS. written in French, containing his notes on birds; but I was not permitted by the owner to do more than glance at it, although he offered it for sale at what seemed to me a very exorbitant price. Fearing to lose a book which might prove of considerable utility in the pursuit of my favourite science, I yielded to his demands and secured the coveted prize, but found upon perusal, amidst a mass of bad grammar, bad spelling, and worse writing, which cost many hours to decipher, that it did not contain so much information as I had reason to anticipate, a good deal of the matter having been copied from other authors.

However, there was some grain among all this chaff; and such facts and details as are considered worth recording are given below with Favier's name attached, and supplemented by my own observations in Morocco. These notes I keep separate from those referring to the Spanish side of the Straits.

This little work is, it may be distinctly understood, not intended to give any special information to scientific ornithologists, but is published with the view of assisting with trustworthy information any amateur collectors who visit Southwestern Europe; and it is hoped especially that it may be useful to officers who, like the writer, may find themselves quartered at Gibraltar. For it admits of little doubt that the study of Natural History will always help to pass away with pleasure many hours that would otherwise be weary and tedious during the time military men may have to " put in " at dear, scorching old " Gib."

There is ample room, for any one with energy, to work out a great deal more information on the birds of the Straits; but it must be remembered that little can be done in hasty visits of two or three months, or by comparing skins secured by assistants, *called* " collectors," who know nothing of the habits of the birds they send to those who employ them, and upon whose veracity even as to locality the employer cannot implicitly depend.

It may be as well to notice such papers &c. as have been published hitherto relating to the ornithology of the district of the Straits. Dismissing the Spanish lists as meagre and full of errors, we commence with the papers written by Lord Lilford in 'The Ibis' for 1865 (p. 166) and 1866 (pp. 173 & 377). In addition to the interesting facts they contain, these essays are the first which give any reliable information on the subject, and lay, as it were, the foundation of all the work that has since been done with regard to Spanish ornithology.

Since then Mr. Howard Saunders has written, also in 'The Ibis' (1871, pp. 54, 205, & 384), a "List of the Birds of Southern Spain," extending as far eastward as Minorca and as far to the north as the fortieth degree of latitude, thus comprising a very large area. He has also contributed some other papers to 'The Ibis' (1869, pp. 170 & 391), which, altered and considerably enlarged, have appeared in 'The Field' under the head of "Ornithological Rambles in Southern Spain."

On the African side, Mr. Tyrwhitt-Drake gave a list of the birds observed by him in Tangier and Eastern Morocco (Ibis, 1867, p. 421); and a "List of Birds seen near Tangier" appeared in 'Naumannia,' but, as far as I can judge, it is only a list, and a very inaccurate one.

Lists of birds, generally speaking, have very few dates appended; the exact localities where a species may be found is seldom indicated; the period of migration is also not often stated. "Breeds plentifully," "appears in winter," "a regular visitant," "abundant in spring,"—such are usually the vague remarks given with each species.

Many of the ornithological papers in periodicals and journals are written up from one or two passing visits, often very short ones; and some of the writers possess a power of vision truly astonishing. They see a bird in the distance, imagine it to belong to a certain species, at last believe it to be such, and end by placing the name in their note-book, to appear in due time in print.

The only way to avoid such errors is never to include any bird in a list except when actually obtained and identified. It often happens, also, that the bird *seen* and included is one which it would be quite impossible to distinguish from another closely allied species without handling them both.

These remarks may, no doubt, appear very invidious; but it is want of accuracy in such matters which renders utterly futile any attempt to make out the distribution of birds.

Local names, often trivial and unimportant, must generally be accepted *cum grano salis*; for, unless long resident and conversant with the language of the country, the compiler is apt to fall into the same class of errors as those of the celebrated Count Smorltork, who would probably have written the English name of the Curlew according to the story told of a gunner in the Eastern Counties, who, when asked by a portly old citizen, "What do you call those birds?" replied, "Bless you, Curlews we *generally* calls 'em; but when we're vexed with 'em, we calls 'em BEGGARS." These vernacular names are most useful, of course, in the case of the more common species, and in Andalucia are, in many instances, of Arabic derivation, relics of the Moorish occupation and of days when under their rule Spain was flourishing, when all that is worth seeing was built, all that is artificially good being remnants of the work of the then industrious Moors. Where are the latter now as a nation?

As a proof of the inaccuracy of local nomenclature, a single name is often applied to several species, sometimes not even belonging to the same genus. Thus *Aguila*, *Aguilucho*, according to the ideas of the individual, may be any of the Diurnal Accipitres, from a Lammergeyer to a Lesser Kestrel; and they are even occasionally used to designate the Raven!!

So *Bujo* applies to *all* Owls, *Culiblanco* to *all* Wheatears, *Chorlito*, the real name of the Golden Plover, is used for various Waders; while *Pitillo*, *Frailecillo*, *Andarios*, *Correrios* are indefinite names applicable to any small Waders and some larger ones. *Pito real* near Gibraltar is *Picus major*, our Great Spotted Woodpecker; near Seville it is *Gecinus Sharpii*, the representative of our English Green Woodpecker (*G. viridis*). *Carpintero* in Central Spain, according to Lord Lilford, is *Picus major*; near Gibraltar it is the Great Titmouse (*Parus major*). *Lavandera*, or "washerwoman," according to localities is either a Wagtail or a Green Sandpiper. *Quebranthuesos*, "bone-breaker," properly applies to the Lammergeyer; but where that species is absent, it is usurped by the *Neophron*.

These, among other instances, prove local names to be only an assistance, and not always to be taken to signify the bird to which they are affixed.

On the other hand, some names are distinctive, as *Abejaruco*, Bee-eater; *Abubilla*, Hoopoe; *Abujeta*, Godwit; *Alcaravan*, Stone-Curlew, &c.

The Moorish Arabic names are for the most part copied from Favier's MS.; but I have quoted none unless corroborated by reference to natives of the localities in which the birds were shot. I will only add further that, as it is a matter of considerable difficulty to express Arabic words by English letters, in this work a mere approximation to the sound is attempted to be given.

In the interest of the sportsman and the amateur collector of specimens, I have endeavoured to give a few hints as to the localities where each may best gratify his tastes; but it would be foreign to my purpose and inapplicable to the limits of this work were I to reproduce any of the varied information which is to be found in the pages of Ford and other guide-books. In the country districts of Spain, and especially throughout Andalucia, nothing ever changes except the Government, which now can hardly be said to exist. The country is still the Spain of Ford, I might almost say of Don Quixote, and will probably remain so for centuries, except as regards the power of the priesthood, which is gradually waning and will doubtless soon cease to exist.

In a general sense, travelling in Morocco is attended with more expense and less comfort than in Spain. The total absence of inns in the former country (except one or two at Tangier and a few coast-towns) renders it necessary for the European traveller to carry about not only a tent but a good deal of extra baggage, cooking-appliances &c., which would otherwise be superfluous. The *impedimenta* are transported on the backs of mules, which may be hired at the rate of one dollar *per diem*. One or two Moors must also be taken to pitch tents, load and unload the pack-animals, and so make themselves generally useful, which they always do. They are usually paid one and sixpence a day each. The only provisions which can be procured at the country villages consist of fowls, eggs, butter, milk, and kuskoo-soo; the latter is a peculiar preparation of flour, and may be considered the national dish of Morocco. It would therefore be advisable for a party travelling in the interior to provide themselves with some tins of preserved meat and vegetables, as well as with whatever wine, spirits, &c. they might require. The most satisfactory way of making an expedition through the country, I have found by experience, is to contract with a respectable Moor, who will usually defray the entire expenses, including hire of baggage, animals, servants, cook, and meals, *exclusive of wine*, at the rate of five dollars a head *per diem*.

As far as I have had opportunities of judging, I have reason to consider travelling in Morocco to be unattended with any danger; and to Englishmen the natives are certainly better inclined than to any other European nation. The late Sultan, however, issued an edict to the effect that he would not hold himself responsible for the life or property of any Christian who at the time of any outrage should be unattended by a Moorish soldier; and I may here state that a mounted soldier to act as a guard can always be procured on application to the Consul of the nation to which the applicant belongs. I should therefore strongly advise any party intending to make an expedition further than ten or fifteen miles from Tangier to provide themselves with this necessary functionary, to whom they must pay one dollar a day. This sum is generally considered to be exorbitant; and it certainly is so in a country where the necessaries of life are far cheaper than in any country in Europe. The real truth, however, is that the Moorish authorities throw as many difficulties as they possibly can in the way of inquisitive European travellers, well knowing that, were the interior thoroughly opened up, the vile system of oppression and misgovernment to which it is and has been for so long a victim, would vanish before the opinion of the civilized world. Although to those who are acquainted with the general system of Moorish rule (or rather misrule) it may not appear strange that such a wretched government should endeavour indirectly to check the influx of civilization, yet what does appear to me most strange is the fact that such a policy should be connived at by the representatives at Tangier of at least some European states; and that this is the case is, I maintain, proved by the said representatives permitting, without murmur or remonstrance, the imposition of an exorbitant tax, in direct violation of the spirit of the last treaty, which secures for the subjects of those states free transit through the empire and the enjoyment of every privilege and right accorded to the Moors themselves.

The vicinity of Tangier is as good a ground for the ornithologist as can be wished anywhere; but it has been a great deal worked up by Olcesse, who succeeded Favier as the naturalist of Tangier. About twelve miles to the south are the lakes of Sharf el Akab, well worth visiting for aquatic birds. The country beyond this to Larache is not good until within the neighbourhood of that town, where there is plenty of both marshy and dry ground, the latter, in places, better-wooded than usual. Near Larache, on the north bank of the river, are the ruins of the ancient Lixus, at or near the spot where Hercules is supposed to have conquered Antæus, the founder of Tangier, which takes its name from his wife Tinga. Very tolerable quarters are to be got in Larache at the house of one Rafael Pitto, a native of Gibraltar, who knows the country very well and is very efficient as a sporting-guide. South of Larache are the lakes of Masharalhaddar and Ras Doura, the latter running for miles southwards in the direction of Rabat. These lakes swarm with every kind of aquatic bird, according to the season; but in the breeding-time the mosquitoes are enough to drive any European away, besides which the nests are so plundered by the Arabs that it is hardly worth while going there for them. Further south than this I have not been, and refer my readers for any information to Mr. Drake's paper in 'The Ibis' (*l. c.*).

Eastward of Tangier, taking the road to Tetuan, there is little or nothing to be done in the way of birds until the latter place is reached, after a long and tedious day's journey; indeed all that part of Morocco which I have visited is very wearisome to travel over, except near Tetuan and Ceuta, where the mountains break the sameness of the route, and where alone any true beauty of scenery is to be found.

Of these hills only those in the immediate vicinity of Tetuan can be visited, owing to the lawless character of the hill tribes and their Mahometan prejudices, and, last but not least, owing also to the exaggerated stories made up to prevent any European from travelling about. In a stream on one of these mountains, to the south of Tetuan, a species of trout is found; but I was unable to get a specimen: they are also probably met with in the other, higher mountains, which are forbidden ground to the European.

The country about Tetuan is alike interesting to the ornithologist and favourable to the sportsman; about Martine are some fine marshes, while beyond Cape Negro, towards Ceuta, is a large, irregularly shaped, shallow laguna, called Esmir, with great masses of rush and sedge, interspersed with tamarisk bushes, separated from the sea by a wide sandbank covered with brushwood; this laguna and marshes are by far the best ground I have seen on either side of the Straits. Proceeding from Esmir, towards Ceuta, the road lies either on the shore or along the usual scrub-covered country till, turning to the left by some Roman ruins, a pass leading up to the Sierra Bullones is entered, when the scenery becomes very fine, the track ascending by the side of a bright clear stream, through bushes sometimes so thick as to completely shut out the sky overhead, at other times passing through heather, in places twenty feet high. The path becomes gradually worse, till the climax is reached in the ascent of a steep hill where the brushwood tears the load off the mules, and with the stones and rocks nearly renders progress impossible. Once, however, at the top, a fair enough road is found leading to the village of Beut, situated in a sort of plateau at an elevation of about 1000 feet, separated from Jebel Moosa by a deep valley, a high range of rocks, and another shallow valley.

About here nothing, ornithologically speaking, is to be seen (excepting a few Choughs) that is not to be met with elsewhere. We found an Eagle nesting on the north face of the range south of Jebel Moosa: the nest was in a most difficult position to get at; and not being able to reach any place near enough from which to shoot the Eagle, we left the eggs as worthless, because unable to identify the bird: however I have little doubt that it was the Golden Eagle. We saw some apes about the rocks; they were rather wild, and lost no time in making their way to the top.

The view from this sierra (Apes' Hill of the English, Jebel Moosa of the Moors, Sierra Bullones of the Spaniards, Abÿla of the ancients) is magnificent, and baffles description, well repaying all the trouble and difficulties of the ascent.

To the south beyond Tetuan lie, half hidden in blue mist, the snow-streaked mountains of the Atlas, stretching far away out of sight, the summit of one vast snowy pile rather to the south-east appearing to be as high, and looking quite as white, as the Sierra Nevada, near Granada, which is also distinctly visible to the north-east; but this African snowy range seems further off. Below to the east, stretched out as if on a ground-plan close to your feet, is Ceuta, with its detestably ugly white-washed Spanish forts and towers, between which and the Tetuan river looms the gloomy headland of Cape Negro. Westward of this are range after range of comparatively low dark hills, rolling away towards Tangier and Cape Spartel, here and there one or two being topped with a few crags and rocks. Far to the west is the Atlantic, leading to the deep-blue Straits, looking, as they separate Europe and Africa, like some mountain-lake.

Tarifa, to the north-west, is clearly visible, as are the crags of the Sierra de San Bartolome, the sandy cliffs of Cape Trafalgar, and the long spit of land on which is the lighthouse; while all the grey, bare, barren-looking Spanish sierras look, with the sun shining on them, as if they lay within a stone's throw.

Gibraltar was shut out from our view, owing to the pleasantry of some Moors, who had rolled large stones down the only path leading to the summit of the highest peak, and so prevented us from ascending. However the view we did get was such as can never be forgotten, and it was long before we descended to continue our hunt for birds.

The tops of these mountains, which are 2600 feet high (the highest part of Jebel Moosa is about 2800 feet), were covered with thousands of violets then in full bloom. The flowers were light-coloured when growing among the stones and waterworn rocks exposed to the sun, dark when shaded and growing among the stunted bushes which were scattered about here and there: their scent was perfect. Very few other flowers were growing on the tops; but most conspicuous among them was the Gibraltar candytuft; and the everlasting palmetto was met with at the very highest places. The base of all these crags or cliffs is approached by a steep slope of small broken rocks, among which grows a very thick jungle of stunted cork- and olive-trees about 15 feet high.

On the north side of the range next to Beut and at the western end of it, at the base of the cliff, is a wide cave, which, at about some fifty or sixty yards from the entrance, branches off into two distinct caves, one going up hill, the other down. My companion ascended the upper one till he heard in the darkness the growling of some beast, probably a lynx or some wild cat; so he returned, and we collected together from outside a lot of dead sticks and rubbish, which we dragged up in the darkness as near the animal as we could judge to be well within range: we then set fire to it, and stood ready for a shot; but it was no use; the brute, whatever it was, only retired further in, growling away more than ever.

The light of the fire proved the cave to be some hundred feet high, gradually narrowing to the top from the bottom, which near the entrance is about 12 feet wide, thus showing it to have been formed by two gigantic rocks or cliffs flung against one another at the time these limestone mountains were thrown up from the bottom of the sea, which in remote ages doubtless flowed over them. On both sides of the Straits, *i. e.* at Gibraltar and Abÿla, these fissures or caves are common in the limestone; but this particular one fairly rivals the well-known St. Michael's Cave at Gibraltar, and had, from the marks of fire, been inhabited at some not very distant period. The floor in places was quite a foot deep with the guano of Rock-Doves (*Columba livia*), numbers of which flew out from the nooks and crannies of the rock.

As far as I could understand from the Moors, who, living near Ceuta, spoke a few words of broken Spanish, there was a story of a Moor having ascended this cave till he came out somewhere at the top of the mountain; be this as it may, there was a fine breezy draught of air blowing downwards, which sent the smoke of our fire towards us till we, instead of the beast for whose benefit it was intended, were nearly suffocated.

Having no means of getting torches to further explore the cave, with heavy hearts we left the unknown animal to growl himself to sleep; the Moors insisted, by the way, that what we heard was a "gin," or evil spirit!

The south-east part of the range of the Sierra Bullones is a different formation, showing symptoms of volcanic action; and I could trace signs of lead-ore and antimony in more than one place. Whether any mines will ever be worked in Morocco is doubtful: there is plenty of ground; but at present it is forbidden to look for minerals by the enlightened and despotic Moorish government.

The track or road from Ceuta to Tetuan, after quitting the mountainous district, passes through the interminable scrub usual to the Mediterranean coast; and bad as are the roads in Spain, this one beats them all in roughness; and, owing to the weary sameness of going up and down hill after hill, the journey seems endless.

En route, however, by the shore, nearly opposite Tarifa, lies, shrouded in large thick bushes, the ruins of Alcazar Leguer, a large old castellated Portuguese fort, built about the beginning of the fifteenth century. Some parts of the walls are in fair condition; but the interior is very much dilapidated, and the whole overgrown with wild olive- and fig-trees, brambles and rubbish, desolation beyond description, its only tenants being Owls and (say the Moors) evil spirits. A covered way, formed by two parallel high walls with banquettes on their tops, runs down to the sea-shore, where it is broken down and blocked up with sand; the ruins show signs of unskilful workmanship, and contrast very unfavourably with those of Roman construction, besides which, from being principally built of soft sandstone, they are much weather-worn when exposed to the rain.

Wherever I have wandered about in Morocco the country is singularly destitute of trees of any size, what few there are being in the santos or graveyards. The consequence of this is, there is no change in the landscape; stunted bushes, rocks, and flat cultivation constitute the general view. Nevertheless the climate is splendid and healthy, perhaps better than that of Andalucia; and one quits it with the regret that such a fine country should in these days of civilization be, as it were, utterly wasted—a land rich beyond most in soil, minerals, and natural advantages of all sorts, within five days of England, remaining without any real government, without roads, bridges, or any means of communication, owing to political necessities abandoned to barbarians and consuls, with both of whom the chief object seems to be to keep the country as much as possible secluded from the prying eyes of Europeans.

CHAPTER II.

The migration of birds, although a most interesting subject, is yet very imperfectly understood, and reliable data from different countries and places are greatly wanted to elucidate it. Without doubt caused by the absence or abundance of food, which in turn is caused by difference of temperature, the passage of birds in these parts begins with most species almost to a day in the spring, usually lasting for about three weeks, though some, as the Hoopoe and the Swallows, are more irregular in their first appearance; and with these the migration lasts throughout a longer period.

Few (indeed hardly any birds) do not migrate or shift their ground to some extent. I can name very few which do not appear to move, viz. Griffon Vulture, Imperial Eagle, Eagle-Owl, Blue Thrush, all the Woodpeckers, Tree-Creeper, Black-headed Warbler, Dartford Warbler, Crested Lark, Chough, Raven, Magpie, Red-legged and Barbary Partridges, and the Andalucian Quail. Generally speaking, it seems to me that in the vernal migration the males are the first to arrive, as with the Wheatears, Nightingales, Night-Herons, Bee-eaters; but this is a theory which requires more confirmation. Some species, as the *Neophron* and most of the Raptores, pass in pairs.

Most of the land-birds pass by day, usually crossing the Straits in the morning. The waders are, as a rule, not seen on passage; so it may be concluded they pass by night, although I have occasionally observed Peewits, Golden Plover, Terns, and Gulls, passing by day.

The autumnal or return migration is less conspicuous than the vernal: and whether the passage is performed by night, or whether birds return by some other route, or whether they pass straight on, not lingering by the way as in spring, is an open question; but during the autumn months passed by me at Gibraltar I failed to notice the passage as in spring, though more than once during the month of August, which I spent at Gibraltar, myself and others distinctly heard Bee-eaters passing south at night, and so conclude other birds may do the same.

We have (*vide* Andersson's 'Birds of Damara Land,' pp. 18-21) an account of the swarms of Hawks which appear there at the time they are absent from Europe and North Africa; so it may be reasonably inferred with regard to one species, *Milvus korschun* (the Black Kite), that some of the vast numbers which pass the Straits of Gibraltar retire in autumn through the tropics to South Africa.

The best site for watching the departure of the vernal migration is at Tangier, where just outside the town the well-known plain called the Marshan, a high piece of ground that in England would be called a common seems to be the starting-point of half the small birds that visit Europe.

Both the vernal and autumnal migrations are generally executed during an easterly wind, or Levanter: at one time I thought that this was essential to the passage; but it appears not to be the case, as whether it be an east or west wind, if it be the time for migration, birds will pass, though they linger longer on the African coast before starting if the wind be westerly; and all the very large flights of Raptores (Kites, Neophrons, Honey-Buzzards, &c.) which I have seen passed with a Levanter. After observing the passage for five springs I am unable to come to any decided opinion, the truth being that as an east wind is the prevalent one, the idea has been started that migration always takes place during that wind. Nevertheless it is an undoubted fact that during the autumnal or southern migration of the Quail in September, they collect in vast numbers on the European side, if there be a west wind, and seem not to be able to pass until it changes to the east; this is so much the case that, if the wind keeps in that quarter during the migration, none hardly are to be seen.

On some occasions the passage of the larger birds of prey is a most wonderful sight; but of all the remarkable flights of any single species, that of the common Crane has been the most noteworthy that has come under my own observation.

On the Andalucian side the number of birds seen even by

the ordinary traveller appears strikingly large, this being, no doubt, in a great measure caused by the quantity which are, for ten months at least out of the year, more or less on migration; that is to say, with the exception of June and July, there is no month in which the passage of birds is not noticeable, June being the only one in which there may be said to be absolutely no migration, as during the month of July Cuckoos and some Bee-eaters return to the south.

Though shooting is hardly a subject within the design of an ornithological brochure like the present, yet it generally happens that an ornithologist is also a sportsman; and therefore a few lines on the subject may be acceptable.

In Morocco no large game is found within reach of the European sportsman, excepting wild pigs, which are only to be obtained by the battue system of driving the jungle with beaters and dogs, sitting for hours waiting for the chance of a shot, a class of amusement dignified by the name of a " boar-hunt;" sometimes, where the country is sufficiently open, the real sport of pig-sticking can be had.

No doubt further in the interior there is other large game; but with the exception of shooting an occasional gazelle and a few pigs, there is no opportunity of using the rifle.

The small-game shooting is very good; the abundance of Barbary Partridges in some districts is miraculous; but when killed they are of little value in a culinary point of view, being more dry and tasteless than the Spanish Redleg (*P. rubra*), now so well known in many parts of England.

The number of Snipe in some seasons is very great, especially at Masharalhaddar, where, and also at Ras Dowra, Larache, Sharf el Akab, Martine near Tetuan, and Esmir near Ceuta, as good snipe- and wild fowl-shooting as may be wished for can be obtained. But it is always uncertain sport, as one day swarms are met with, and perhaps on the next day hardly any are to be found. The absence of roads and bridges renders the country in wet weather at times almost impossible to travel over, the tracks becoming a succession of mudholes, and the rivers impassable torrents. This, added to the unpleasant certainty of living under canvas during rainy weather, is a great drawback to winter shooting.

Another, in my opinion insuperable objection to shooting in Morocco is, that if any great quantity of game be bagged, it has to be thrown away, as, unless within twenty miles or so of Tangier, it is useless. The Moors, being Mahometans, will not eat any thing killed by a Christian or infidel; and killing for the mere sake of slaughter does not come within the creed of a real sportsman. In Spain all this is very different, as any one and every one is only too glad to accept of the surplus game.

In almost all parts of Morocco rabbits abound; and hares are in places plentiful. Woodcocks are sometimes tolerably abundant; Quails, of course, are in swarms during migration; and there are a great number of Little Bustard.

Shooting in Andalucia is far more satisfactory and pleasant sport than on the African side. In the first place, accommodation can always be had in a house of some sort: the ventas are much to be preferred to *cortijos*, or farm-houses, as the latter usually swarm with fleas; but by taking your own blankets and a camping-palliasse, which can be refilled at each resting-place with chopped straw, one can generally by the aid of a liberal use of flea-powder* manage to cheat the vermin of their nocturnal banquet. It is almost absolutely necessary to take this powder with one, as sleep in some of the dens which I have been in would have been impossible without using it. Another most necessary item is an india-rubber flexible bath, as it is seldom a "*lebrillo*" or large earthenware pan big enough to wash in can be obtained.

In addition to the shelter to be got in Andalucia there are some roads; and bad as they may be, they do afford means of communication; and there are bridges, though not always placed in the right situations; for in places you see a bridge built across a gully without any road on either side of it, and others where the stream has quitted its old course for a new one—single instances out of the many thousand strange and wondrous *cosas de España*.

* This vegetable powder is made from a species of Fever-few (*Pyrethrum*), and is quite innocuous except to insects; many other plants of the Radiatæ group are equally offensive to insects.

The large game is more varied and plentiful in Andalucia than in Morocco. In most of the wooded valleys of the sierras, near Gibraltar, there are a good many roe-deer (*corzo*) and a few wild pigs; in some of the high sierras near Ronda, Ubrique and in the Sierra Nevada the Spanish Ibex is sparingly found; but it is extremely difficult to get them without organizing a regular drive or *batida*—a very expensive affair, requiring a party of not less than eight guns, who must take tents, cooks, &c. up into the mountains; and then, if successful, as far as sport is concerned it is hardly worth while sitting for several hours behind a stone, nine times out of ten without even seeing an ibex. It is almost impossible to stalk them, as they lie hidden in the thick stunted fir and other scrub which is scattered in large patches on the mountain-sides, and are so wary that you cannot come suddenly on them like the roe-deer. However, in an ibex-shooting expedition, one is amply repaid by the magnificent scenery and the novelty of the affair; but as far as shooting goes it is a failure, and at the lowest calculation every ibex killed by a Gibraltar party costs over £100.

Ibex drop their young about the end of April; on one occasion a shooting expedition with which I was present succeeded in getting two, both of which I sent home to the Zoological Gardens; but unfortunately they did not long survive. In the Sierra Morena, near Palma, a little to the west of Cordova, are red deer strictly preserved and well pastured; the "heads" of the stags are very fine, which is not the case with those of the Coto Doñana, near San Lucar de Barrameda. All these, however, being wood-frequenting deer, the antlers do not branch out very widely, most of the heads being rather narrow. It is in small-game shooting that Andalucia excels, though it is in no way equal to that of the countries in the east of the Mediterranean. Foremost, both in numbers and sport, is the Snipe-shooting; for in some seasons, about November and December, if the weather has been dry, it is equal to any that can be obtained; but all depends upon the weather, which, if wet, causes the birds to disperse over the whole country, while if it be dry they remain in the sotos or marshes, and when flushed return almost immediately. Some of the best sport I have had with them was by waiting in favourite ground while they kept coming in, flying high up over head, and then swooping down and pitching within a few yards. I have known fifty couple bagged in a day by one gun, thirty or twenty-eight couple a gun per day being often obtained. The proportion of Jack-snipe is about the same as in England, and they keep to the most wet and muddy spots. Snipe, as a rule, in Andalucia are far more wild than in other countries, which is no doubt caused by the nature of the marshes, which, often quite dry at the end of summer, are in winter regular lakes, only at their edges affording any resting-places for the Snipe, the cover being usually thin and bare.

There are many acres of ground flooded with water, from about six inches to a foot in depth, the whole dotted over with tussocks standing an inch or two above the water, and about a foot apart from each other. This tussocky ground is most difficult both to walk over and shoot on, as the tufts are not broad enough to stand on with both feet, and these slippery lumps of mud and grass standing above the water enable the Snipe to see a long distance, and cause them to rise very wildly; while they also have a most provoking habit of flying up just as you are trying to balance yourself on one of the tussocks. The result, if you fire, is most probably a miss, and down you slip into the water, lucky if on your legs and not on your knees or, as happened to me more than once, on your face. There is, however, one point in favour of all these sotos: they have a firm bottom, the mud is never deep, and there are no quaking bogs or dangerous morasses as in Ireland. A retriever, it is almost needless to add, is perfectly indispensable for this sport, saving (in addition to many birds that would otherwise be lost) much time and the bad temper which results from not being able to find birds that have fallen. Snipe in Andalucia are very seldom seen in lots together or in wisps, though occasionally in very wet stormy weather small wisps appear. The best localities which I have visited in Andalucia are the marshes near the edges of the Marisma, or delta of the Guadalquivir, below Seville, especially just beyond Coria del Rio, and near the Coto del Rey and the Coto Doñana; one spot near the Palacio of the former place, las Carnicerias, is excellent. At Casa Vieja or, more properly speaking, Casas Viejas, some forty miles from Gibraltar, is very good ground, particularly in the first part of the season; there are also good marshes near Vejer. Late in the season, near Taivilla and Tapatanilla, on the road from Tarifa to Vejer, at times Snipe are also to be found very plentifully.

The wildfowl- or duck-shooting in dry seasons is very fair in the early part of the winter, before the lagoons and rivers are filled up by the rains, there being then very few wet spots, and the birds crowd together in the small pools which remain between the high banks of the river-beds, and can be easily approached; but later on, when these streams are brimful or, rather, overflow their banks, and when the lagunas are sheets of water without rushes or cover of any sort at the edges, it is almost impossible to shoot ducks by day except by making "hides" with sticks and stones, and sending some one round and trying to have them driven over you. At flight sometimes very fair sport is to be had for one or two nights; but after that the fowl know the place, and either come very late or avoid it altogether. For flight-shooting a good retriever is necessary; for it is, in the dark, impossible to find the spoil; and if left till morning, the Marsh-Harriers are at them by break of day, leaving nothing but bones and feathers. To my mind there is very great charm in flight-shooting, though sometimes it is hot work while it lasts; for it is over in about half an hour, and you cannot, even with a breech-loader, load quickly enough. It requires, too, considerable skill in judging the distance, and sharpness of vision in being able to catch a glimpse of the ducks as they pass over. It is a great help if you can place yourself so that you face the west, and thus get the birds in the evening light, when they can be seen coming a very long way off; but if they come from the eastward, and you are obliged to face that way, they never show till close on you; and the frogs make such a noise, croaking, that you cannot, as in England, hear the sound of the duck's wings.

Immense numbers of Wild Geese are found in the winter months about the Laguna de la Janda, and below Seville, in the marshes of the Guadalquivir. They are of course very difficult to "get at;" but as they pass the day on the ground at the edge of the water, and always have certain favourite spots to which they resort, they are to be got by digging or making "hides" at the places they most frequent. In the morning, at sunrise, they collect on the water, in some places in hundreds, and swim about feeding for an hour or two on some substance which they pick up from the bottom of the shallow water; after this they disperse and take to the shore, where, if left undisturbed, they pass the day sleeping and pluming themselves. There is one of these goose-haunts near the Palacio of the Coto del Rey, a little to the south-east of it. One morning in January, having the day previously made a hide among some tufts of rushes, I went and laid up before sunrise to await the geese, which arrived by degrees in flight after flight, till there must have been within a mile of me, at the lowest computation, between three and four thousand; I shall never forget the sight, and I lay concealed watching them for at least two hours. I could not distinguish amongst them more than one lot of about a dozen Bean-Geese (*Anser segetum*); the remainder were all Grey Lag (*A. ferus*). Some hundreds were within about a hundred yards of me, and it was very amusing to see them feeding, fighting, and playing with one another; but somehow they were evidently suspicious of the patch of spiky rushes in which I was lying flat in the slight hole which I had made between two tufts of rushes and covered over with others dug up by the roots, and arranged so as to look as if growing. Unable to turn on my side or move in the least, I was so cramped that it was all I could do to remain there; but after a time a large lot of geese began to set in towards my position, and in a few minutes more I should have had a good family shot. I had plenty of chances of firing, but could not have got more than a couple; besides which I wished to watch them, so waited in hopes of a good flock coming close to me, when, alas! I heard cries of alarm from the birds furthest away on my right, and after a minute or two they began to fly up, and I could see against the sky a man riding towards them. The geese in front of me all pricked up their heads and were getting ready to be off; so I was obliged to jump and send both barrels at them as my only chance; and by good luck, or rather thanks to the large shot, I killed two, not enough to recompense one for lying cramped up for so long; but still I was more than repaid by the sight of so many wild geese close to me, and being able to watch their movements. Any one who would take the trouble to try punt-shooting with a big gun below Seville and on the Laguna de la Janda might make some wonderful bags, as the enormous quantities of Geese, Wigeon, and other ducks can only be approached with the aid of a punt. When near the edge of water you can always approach Ducks with a stalking-horse; and Geese will allow this on their first arrival, but soon become too wary.

Golden Plover are extremely abundant in vast flocks from November to March. On their first arrival they are not so wild as afterwards. They can always be "got at" with a stalking-horse; but as good a plan to shoot them as any is to stand still in some place which they frequent on a windy day, when they will often fly within a few yards. Peewits are numerous, but not worth shooting, as is the case with Curlews; but the latter are, as elsewhere, much too wary to allow themselves to be shot, and during the whole time I was in Andalucia I never but once had the chance of killing one.

Woodcocks in some seasons are numerous; but five or six couple in a day is a very good bag, very different from Albanian shooting. Partridges (*C. rubra*) are not worth the trouble of going after, either for sport or for the table; in some places there are a good number, but not near Gibraltar; they are the chief object of sport with the Spaniards, who kill them at all seasons, even shooting them from the nest. Quails are, during the *entrada* or autumnal migration, so extremely abundant that, if there has been a westerly wind for a few days in September when they are on passage, there is really no limit to the number that may be shot. About Tarifa at that season every gun-possessing man and boy turns out with every cur dog in the town, and, regardless of each other, they fire in all directions, so that it is a service of danger to go out near them. If the wind during their passage remains in the east, the Quails pass on, and little or no sport is to be had with them. A west wind detains them and prevents their passing the Straits, though it does not seem to retard their migration by land.

The remaining small game to be noticed in Andalucia are Bustards, both the Great and Little Bustard, hares, and rabbits. The Great Bustard is only to be got with any certainty by driving. The Little Bustard, more wary still, is only to be shot in the end of July and in August during the extreme heat of the day, though sometimes they can be driven over a gun by getting under the bank of a river or such like shelter, and sending a man round to put them up; but on rising they usually mount up very high, in this respect differing from the Great Bustard, which hardly ever flies high enough to be out of shot if you are directly underneath. Hares (*Lepus mediterraneus*) are a much smaller species than in England, about the size of a good average English rabbit, not very abundant anywhere and frequenting open flat and cultivated districts, never being found among woods or hilly ground. Rabbits, of course, are abundant but very small, rather less in size than the New-Forest rabbit, which is the most diminutive race in England. A shooting-license, easily obtainable through the aid of any British Consul, is requisite in Spain; and though seldom asked for, it is better to have one. The form and cost of one varies according to the Government, and therefore is seldom alike two years in succession. Lately there have been two licenses—one to carry a gun, the other to kill game, though what " game " is not defined.

I here mention some notes of the Mammalia of Andalucia, with their local names, which may be useful to the sportsman. Of course there are other species of small Mammalia to be found, especially among the bats; with the names of the latter I have been kindly assisted by Lord Lilford, who has

MAMMALIA. 23

personally obtained them all in Andalucia. Those marked with an asterisk I have obtained myself or seen "in the flesh."

Greater Horse-shoe Bat	Rhinolophus ferrum-equinum.
*	Rhinolophus euryale.
Lesser Horse-shoe Bat..	R. bihastatus.
	Dysopes rueppellii.
Barbastelle	Barbastellus communis.
*Noctule	Vespertilio noctula.
*Mouse-coloured Bat ..	V. murinus.
*Schreiber's Bat	V. schreiberi.
Long-eared Bat.	V. auritus.

V. schreiberi, V. murinus, and *R. euryale* are found in caves near Casa Vieja—the two former species in countless numbers, the dung at the bottom of the caves being from four to five feet in depth. The Spanish name for all is *Murcielago*.

		Spanish names.
*Hedgehog	Erinaceus europæus.	Erizo.
*Shrew	Sorex araneus.	Musaraña.
*Mole	Talpa europæa.	Topo.
*Badger...........	Meles taxus.	Tejon.
*Common Marten Cat..	Mustela foina.	Foina.
*Polecat	M. putorius.	Turon.
*Weasel	M. vulgaris.	Comadreja.
*Otter	Lutra vulgaris.	Nutra or Nutria.
*Genet	Viverra genetta.	Gineta.
*Ichneumon	Herpestes widdringtonii.	Melon, Meloncillo.
*Wild Cat	Felis catus.	Gato montés.
*Spanish or Spotted Lynx	Felis pardina.	Gato clavo, Gato cerval.
*Wolf	Canis lupus.	Lobo.
*Fox	C. vulpes, var. melanogaster.	Zorro.
Squirrel	Sciurus vulgaris.	Ardilla.
*Fat Dormouse......	Myoxus glis.	Liron campestre.
	M. nitela.	Raton careto.
*Dormouse	M. avellanarius.	Liron de los Avellanos.
*Brown Rat	Mus decumanus.	Rata.
*Mouse	M. musculus.	Raton.
*Black Rat	M. rattus.	Rata negro.
*Long-tailed Field-Mouse	M. sylvaticus.	Raton de campo.
*Water-Rat	Arvicola amphibius.	Rata de agua.
*Field-Mouse	A. agrestis.	Topino.
*Hare	Lepus mediterraneus.	Liebre.
*Rabbit	L. cuniculus.	Conejo.
*Wild Pig..........	Sus scrofa.	Jabali, Jabalina.
*Red Deer.........	Cervus elaphus.	Ciervo.
*Fallow Deer	C. dama.	Gamo, Paleto.
*Roe Deer.........	C. capreolus.	Corzo.
*Ibex	Capra hispanica.	Cabra montés.

In this book I have endeavoured to name with each species of bird some definite locality where they may be found, which is rather necessary, as certainly on the Spanish side of the Straits birds are very locally distributed, more so than in most countries I have seen. It is difficult to surmise the cause of this, as precisely similar tracts of country within no very great distance of each other are not always frequented by the same birds. On the Spanish side, without doubt, the most common bird as regards numbers is the Goldfinch (*Carduelis elegans*), and the most universally distributed the Stonechat. The number of birds of prey is very great, chiefly feeding on rabbits, rats, mice, reptiles, and insects; they are very useful, and as the ground-breeding birds suffer much in the nesting-season from snakes and lizards, those birds of prey which feed mostly on these enemies of the smaller birds render their lesser feathered brethren valuable protection. The number of small birds, especially during the season of migration, is sure to be noticed even by the most unobservant. Immense quantities of Larks, Finches, and even some of the Warblers are brought into the markets; but as a Spaniard seldom shoots at such small fry, they are chiefly netted, caught at night with a lantern and bell, or snared with bird-lime (*liga*).

The best localities for an ornithologist living at Gibraltar to obtain specimens or watch migration is the country west of an imaginary line drawn due north from Gibraltar as far as the latitude of Seville. Within this district, part of which is given in the Map attached to this volume, as much can be done as is possible in three or four months' time; and the district is large enough to require three or four years to work it out thoroughly.

In the immediate vicinity of Gibraltar (or el Peñon, as the Spaniards call it), the cork-wood of Almoraima and the level ground, mud-flats, and old salinas "between the rivers" on the way to Algeciraz offer to the collector capital ground for work. In the cork-wood particularly, several birds are found breeding which do not seem to nest elsewhere. The ground north-east of Gibraltar is to a great extent covered with scrub and brushwood; and little is to be done in the bird line in that direction.

The sierras being too far distant, cannot be worked from Gibraltar; it is necessary to go to Algeciraz, Faginas, Pulverilla, or some *cortijo* near the hills you wish to work. Very deceptive in appearance, looking quite low and easy to ascend, it takes three or four hours to reach their tops, which, bare, rugged, and wild beyond description, are alone worth visiting for the view, which, always grand, on a clear day is magnificent, that from the Peñon del Fraile at the west of Algeciraz being the finest. From these mountains run down numerous wooded valleys (*gargantas*) clothed with cork and oak trees, many of very large size, though badly mutilated by being lopped by charcoal-burners. The rocky streams which flow down these valleys are fringed with rhododendron, arbutus, holly, hawthorn, laurestinus, bay, myrtle, giant heather, cistus, and many sorts of ferns, conspicuous amongst them being the *Osmunda* and maiden-hair, while here and there is an occasional *Caladium* with its huge leaves reminding one in shape of elephants' ears: the leaves of this plant, called *hojas de llama*, are much used by the country people as a medicine for fevers; many of the rocks and all the trunks of the cork trees are festooned with hare's-foot fern (*calaguala*), also used medicinally.

In spring these ravines are, from their natural beauty and the colour of these various shrubs and flowers, so picturesque that one cannot help lingering about them merely to admire the charming scenery, becoming apt to forget the birds for which one is in search. These places are seldom visited by an Englishman, only by stray smugglers, goatherds, and charcoal-burners; and every pass, hill, valley, in fact every well-marked situation, has its name, many as familiar to me as the streets of London.

Those valleys most worth visiting near Gibraltar are the Garganta del Capitan, to the north-west of Algeciraz, on the way to Ojen by the mountain-path of la Trocha, which is within easy distance (five or six miles) of Algeciraz.

The valley of the Guadalmalcil, halfway on the road

between Tarifa and Algeciraz, is also very beautiful; but the Garganta del Helecho (Valley of the Ferns), south-west of Pulverilla, is perhaps the best for shrubs, flowers, and ferns. The "Waterfall" valley, near Algeciraz (la Garganta del Aguila), is tamer than any; but above the cascades or waterfalls it improves on acquaintance. This ravine, however, is well known to every one who has been at Gibraltar, as the regular rendezvous for picnics, the very name of which is enough to destroy any merits that the scenery may possess. Towards Tarifa and beyond, on the road to Vejer, the country is not so pretty, opening out near Fasinas to the vega of the Laguna de la Janda; thence, cultivated ground, or *campiña*, stretches away to Medina Sidonia and on to Jerez. On the right and left of this road, however, are three isolated rocky ranges—those of la Sierra de San Bartolomé and la Sierra de la Plata being to the left, that of la Sierra Enmedia to the right; these ranges are the breeding-places of Griffon Vultures and other rock-breeding birds, and are well worthy of a visit.

I here give the names of a few of the rocky cliffs which should be visited by those who wish to see such places:—la Laja de la Zarga and la Silla del Papa, in the Sierra de Plata; la Laja del Sicar, to the east of and near Taivilla; Piedra de Paz, near Paterna; la Laja de los Pajaros, los Jolluelos, and la Laja de Peñarroyo, near Casa Vieja.

There are also magnificent cliffs in the Sierra de las Cabras, east of Alcala de los Garzules, and hundreds of others which I saw but could not find time to visit. I did not care to send "collectors" to bring eggs without the birds to which they belonged; or, as is often the case with these worthies, they would have brought eggs with birds to which they did not belong, and, with unblushing effrontery, sworn perhaps, as I have known them do, that a Turkey's egg was taken by them in a high cliff, and belongs to an "Aguila de las rocas."

I hope that this book may not be the cause of the useless or unnecessary destruction of any bird, and especially that dealers may not profit thereby. All I have mentioned is intended for the benefit of true ornithologists, and not for those

(I am sorry to say there are some) who make a regular trade under the name of Ornithology, and are never satisfied unless killing or having killed as many rare birds as possible.

It will be seen that there is some sport to be had in Andalucia; and the shooting has the charm of a varied bag, and the freedom to wander where you like, as a rule; added to which it is necessary to work for your game, which, in my idea, adds much to the pleasure of the sport. The climate, too, is all that can be wished, especially in spring, when there is something most exhilarating in the air; but in autumn, until October, it is too warm to go out with pleasure, and the sun-baked, tawny, dusty, thirsty-looking country has lost all the beauty of its flowers and has none of the verdure of spring. To see Andalucia, it should be visited in March, April, and May, in order to thoroughly appreciate both the climate and the scenery.

Another hint which I would fain give is to be as civil as possible, and conform to the customs of the country. The Andalucian peasant, courteous and polite, is at heart a *caballero*, and very different from the inhabitants of the towns; at the same time he is proud and independent, and, to humour him, must be treated with on terms of equality. Above all things remember that it is no use attempting to hurry in Spain, where patience is more severely taxed than in any other country, and where *no corre priesa* is the order of the day. Certainly the best cure for impatience is to pass a few months among Spaniards.

I must here bring this Introductory Chapter to an end, at the same time apologizing for its shortcomings in the fact that it is the concoction of one who detests pen, ink, and paper, and who is more at home with the gun, rifle, or fishing-rod; so, in the manner of the country which to me has so many charms, let me conclude with the farewell and time-honoured salutation, *Vaya Vd. con Dios.*

BIRDS.

Order ACCIPITRES*.

Family VULTURIDÆ.

1. VULTUR MONACHUS, Linn. Black Vulture.

Spanish. Buitre negro.

This Vulture is mentioned by Favier as having once occurred near Tangier; and there is a specimen in the Norwich Museum from that locality, perhaps the identical bird. It is probably not so rare in Morocco as Favier implies. On the Spanish side of the Straits it is frequently to be seen in winter and early spring, though not nearly so abundant as the Griffon; it is more common near Seville than Gibraltar. Some breed in Andalucia, as I discovered one nest by watching the birds building or, rather, repairing it; for on examination it appeared to be an old nest, probably a Stork's, and was a vast pile of sticks placed on an alder tree, about fifteen feet from the ground, in the midst of the thick jungle of the Soto Malabrigo, near Casa Vieja. This place is almost impenetrable, surrounded by open marsh, and is formed of a mass of huge tussocks placed far apart, on which grow widespreading sallows and brambles well interlaced. The space between these tussocks is covered with rushes and sedges, growing in mud and water, in places up to the waist. In my first expedition to the nest it took me more than half an hour to reach the tree, a distance of only about a hundred and fifty yards from the edge of this paradise of Water-Rails and Aquatic Warblers. Upon climbing the tree it was very difficult to see into the nest, as it so overhung, owing to the great breadth; and, alas! there was no egg, not even any lining.

* The nomenclature of the Birds of Prey is taken from Mr. Bowdler Sharpe's recently published 'Catalogue of Birds' (vol. i.).

A few days after, on the 26th of February, I again examined the nest, only to find it lined with wool and a few rushes. Muster-day at Gibraltar, on the 28th, compelled our return to the Rock; so I engaged a man to take the nest and bring the egg to Gibraltar, which he never did, probably not liking the journey through the swampy jungle. The following year this nest was not used by any birds; but in 1874 a pair of Eagles (apparently *Aquila adalberti*) took possession, rebuilding and lining it with fresh green boughs. This was early in March; and with persistent bad luck, on our return there in April, my friend found nothing in the nest, although the Eagles were about. I imagine that they had been robbed of their eggs, or else had deserted it owing to too frequent examination.

The Black Vulture is said to nest near Utrera; but upon inquiry I could not ascertain such to be the case; they appear to go further north to breed, as Lord Lilford found them nesting towards Madrid, and in one season received no less than some seventy eggs. More solitary in habits than the Griffon, and unlike that Vulture, they build in trees and not in colonies—laying only one egg, about the beginning of April.

This Vulture is to be recognized when on the wing, within a short distance, by its dark appearance. The immature birds are very dark-coloured, becoming lighter with age, till they attain the adult plumage. The bare skin about the head and neck is of a *pale bluish* colour.

2. GYPS HISPANIOLENSIS, Sharpe. Spanish Griffon Vulture.

Gyps fulvus auct., ex Hispaniâ.

Moorish. Enisser. *Spanish.* Buitre, Pajaraco.

"This Vulture occurs commonly near Tangier, both as a resident and on passage, and is often seen feeding in company with the *Neophron* on the same carcass."—*Favier.*

I did not see many Griffon Vultures in Morocco, but there were a few pair about Jebel Moosa in April. Near Gibraltar they are very plentiful, nesting in colonies, not exceeding thirty-five pairs, in holes or, rather, small caves in the perpendicular crags or "lajas," which are found in many of the Sierras. The most important breeding-places near to Gibraltar are the Sierra de San Bartolomé, the Sierra de Plata, el Organo in the Sierra Enmedia, and la Laja del Sicar, all near Taivilla. One egg only is the usual complement; and they lay about the 20th of February. Should the first egg be taken, it seems that they lay again about the 15th of April. Of course it is impossible to prove this; but eggs were laid at that time in nests which had been robbed in February. The egg is usually white, but is occasionally marked with buff-coloured blotches, the nest being sometimes merely three or four bits of green bushes laid on the rock.

It was a fine sight to see thirty or more of these gigantic birds fly out at once with a rushing noisy flight from their nests, which they do if one fires a shot at the bottom of the cliff in which they breed; and this is the only method of finding the exact position of their nests, as well as those of other rock-nesting birds, though later on each large crevice or hole where there is a nest is plainly visible, owing to the dung which covers the face of the rock below, looking as if a bucket of whitewash had been poured out of the cave. Vultures in Andalucia are far more wary than in other countries in which I have seen them, except of course during the breeding-season.

How the numbers which inhabit Andalucia at times find sufficient to eat is a puzzle to me; they must be able to fast for some days, or else travel immense distances for their food, as in the winter and spring it is unusual to see dead animals about; but in the hot parching summer months vast quantities of cattle die of thirst and want of pasture. A bull-fight is a sort of harvest to Vultures, which flock in great numbers to revel on the carcasses of the unfortunate horses that have been so cruelly killed.

The Griffon-Vulture may be distinguished on the wing by its light colour when within reasonable distance, and on closer examination, by the head and neck being covered with white down as well as by the ruff of white feathers. The irides of the adult are light yellow; those of the immature are hazel. The Spanish Griffon has been separated by Mr. Sharpe, on account of its smaller size and tawny colour.

The Sociable Vulture, *Otogyps auricularis* (Daud.), is probably to be found in the southern part of Morocco, and is to be distinguished by the naked *pink* or *flesh-coloured* neck.

3. NEOPHRON PERCNOPTERUS (Linn.). Egyptian Vulture.

Moorish. Rekhama. *Spanish*. Monigero, Relijero, Alimocha, Pernetero, and, away from the Sierras, Quebranthuesos.

"Appears near Tangier in flocks during migration, some remaining to nest in the vicinity, awaiting the return of the autumnal migration, to winter probably in the interior of Africa. Those which pass over to Europe cross from February to April, returning in August and September. They nest on rocks in April, generally laying two eggs, sometimes only one. These have a rough surface, and vary in shape. Sometimes there is an interval of two or three days in the hatching of eggs in the same nest. Fifty four eggs have passed through my hands."—*Favier*.

Near Gibraltar Neophrons, during their stay, are abundantly distributed. Two pair nest annually on "the Rock," going by the name of "Rock-Eagle" among those who would call a Buzzard a Bustard, and *vice versâ*. One nest is below O'Hara's Tower, the other below the Rock gun on the North Front.

Many pass northwards at the end of February, the 23rd being the earliest date on which I saw one; but the greater number, many hundreds, almost always in pairs, pass during March. On the 21st and 24th of that month in 1872, great quantities crossed at the same time as flights of *Nisaetus pennatus, Circaetus gallicus, Buteo vulgaris, Milvus ictinus,* and *Milvus korschun*.

The Neophron usually begins to lay about the 1st of April. The nest is often easily accessible from below, and, placed on a ledge of some overhung rock, generally at the top of a sierra, is composed of a few dead sticks, always lined with wool, rags, and rubbish. I found about a pound of tow in one nest and the sleeve of an old coat; it also contains usually a lot of bones, and stinks horribly.

I have known a third egg laid in a nest from which one had been abstracted, one having been left; but whether the third egg was laid by the same bird is of course " not proven." Two eggs seem to be the usual number, the pair being generally alike; but those from different nests vary very much, some being blackish brown, others quite light, some almost round, others elongated. This Vulture is probably the foulest-feeding bird that lives, and is very omnivorous, devouring any animal substance, even all sorts of excrement; nothing comes amiss to it. Sometimes I have seen them feeding on the sea-shore on dead fish thrown up by the tide.

The adult birds are white, with black wings.

The immature plumage is light brown, with darker wings; and, judging from birds which I have seen kept in captivity, they take three years to attain the adult plumage, though probably in a wild state they would not remain so long in their immature dress.

Family FALCONIDÆ.

4. CIRCUS CYANEUS, Linn. Hen-Harrier.

Moorish. Bou hasin (Father of beauty). *Spanish.* Cenizo.

According to Favier this "is the least common of the Harriers near Tangier, being seldom met with."

On the Spanish side of the Straits, though a resident bird, the Hen-Harrier is most frequently seen in winter; but their numbers fluctuate greatly. I observed more in the winter of 1871–72 than at any other time, particularly about Casa Vieja, seldom, however, coming across an old male.

The adults may be distinguished from *C. macrurus* by the pure white rump. The adult male, as in the two next species, is of a grey colour above; the female is undistinguishable on the wing from that of *C. macrurus*, both having and showing conspicuously the white upper tail-coverts.

5. CIRCUS PYGARGUS (L.). Montagu's Harrier.

Circus cineraceus auct.

" This Harrier passes to Europe in March and April; but some remain to breed near Tangier, where they are nearly as common as the Marsh-Harrier, being seen during passage on all sides in pairs. They nest on the ground, laying five eggs, which vary much in shape, the colour being bluish white, marked with spots of clear blue, which, after the egg is blown, turn yellowish."—*Favier*.

Near Gibraltar, Montagu's Harrier is not often met with; near Seville they are very common, and dark specimens, some of them complete melanisms, are frequently procured. Near Lixus, in Morocco, at the end of April, I found a regular colony: there must have been fifteen or twenty pair on a salt marsh across the river. I had no time to go round and examine the ground, and could not cross the river at that place; but we could see with my telescope the hen birds sitting dotted about the marsh. The males took a particular line across our side of the river; so I shot three for identification.

The adult males are to be distinguished by the dusky black bars on the secondaries. It is likewise a smaller species than *C. macrurus*; and the wings are longer in proportion to the body than in other European Harriers.

6. CIRCUS MACRURUS (Gm.). Pale-chested Harrier.

Favier states that this species occurs on passage in the environs of Tangier in April. In the Norwich Museum there is a specimen labelled "Tangier."

On the Spanish side it is not uncommon in spring near Seville. Lord Lilford was the first to obtain it, in the spring of 1872.

It is a rather smaller bird than *C. cyaneus*, and is more affined to *C. pygargus*; the male is easily distinguishable by the white rump or upper tail-coverts being marked with grey spots; these markings, in a less degree, are always also visible in the females.

7. CIRCUS ÆRUGINOSUS, Linn. The Marsh-Harrier.

Moorish. Hedia (*Favier*). *Spanish.* Aguilucho, Rapiña.

"The most common of the Harriers in Morocco, this bird is both resident and migratory in the vicinity of Tangier. Those which migrate, pass to Europe in February and March, returning in September and October. They commence to breed late in March. Their eggs differ very much in shape, being sometimes round, sometimes elongated."—*Favier*.

In Andalucia, as in Morocco, over all low wet ground, the Marsh-Harrier is to be seen in vast numbers, particularly in winter. Great quantities remain to breed, sometimes as many as twenty nests being within three hundred yards of one another. The latter, loosely constructed with dead sedges, vary much in size and depth, and are usually placed amidst rushes in swamps, but sometimes on the ground among brambles and low brushwood, always near water, though occasionally far from marshes. They begin to lay about the end of March, and at that time fly up to a great height, playing about, and continually uttering their wailing cry. The eggs are bluish white, and usually four or five in number; they certainly vary in size and shape, and are often much stained. Like the eggs of all the Harriers that I am acquainted with, and many others of the Accipitres, when blown and held up to the light they show a bluish tinge. I once found a nest containing only one egg, nearly ready to hatch, and saw another with six eggs (three quite fresh and the other three hard sat on). I believe that, if the first set of eggs be taken, they lay again in a fresh nest, as I found sets of fresh eggs as late as the 2nd of May.

The Marsh-Harrier is a perfect pest to the sportsman, as, slowly hunting along in front, it puts up every snipe and duck that lies in its course, making them unsettled and wild. I have repeatedly seen them flush Little Bustards; but these merely flew fifty yards to the right or left out of the Harriers' line of flight, and settled down again.

Cowardly and ignoble, they are the terror of all the poultry which are in their districts, continually carrying off chickens, and, like other Harriers, are most terribly destructive to the eggs and young of all birds.

On account of these propensities, I never let off a Marsh-Harrier unless it spoiled sport to fire at one. Sometimes, when at Casa Vieja and the snipe were scarce, we used to lie up in the line of the Harriers' flight to their roosting-places; for they always take the same course, and come evening after evening within five minutes of the same time. Upon one occasion a friend and myself killed eleven, and during that visit accounted for over twenty. I also upon every possible opportunity destroyed the nest and shot the old ones; but it was the labour of Sisyphus; for others immediately appeared. However, there was a visible diminution of their numbers at Casa Vieja. I never saw rats in their nests or crops, and believe they have not the courage to kill them: small snakes, frogs, wounded birds, eggs, and nestlings unable to fly form the main part of their prey. I have seen the Marsh-Harrier hawking over the sea about two hundred yards from the shore, where there was shallow water, but could not see what they were taking.

The very old males have the wings and tail ash-grey; when flying in the sun, these parts appear almost white. Females shot at the nest and elsewhere *never* appear to attain this plumage. They are much lighter in colour on the body than the immature birds, which usually have the top of the head and shoulders yellowish white, the rest of the plumage being very dark brown; but they vary much in depth of colour. Some young birds have the head quite white; in others it is the same colour as the body; and I could not ascertain this to be a difference of sex.

I do not know whether it is always the case with the Harriers; but, as far as my observation goes with regard to the Marsh-Harrier, it seems that the males do not sit, as I have shot and seen shot many from the nest, but never saw a male killed flying off the eggs, and have noticed that the males only leave the nesting-places to hunt for prey. I have also observed the same fact with Montagu's Harrier.

8. MELIERAX POLYZONUS, Rüpp. Many-banded Hawk.

Obtained by Mr. Drake at Mogador (Ibis, 1869, p. 153). The Zoological Gardens also had one living a short time back, which came direct from Morocco, and Lord Lilford has recently received one from Mogador. The legs and cere are vermilion.

9. ASTUR PALUMBARIUS, Linn. The Goshawk.

Moorish. El boz (*Favier*). *Spanish.* Azor, Gavilan.

"This Hawk is resident near Tangier, and is frequently seen during passage; but they are rarely met with in winter. They pass northwards in April; those which breed, nest in May. The eggs are pure bluish white, often much stained with yellow. The young are so fierce that sometimes those in the same nest will kill and eat one another."—*Favier.*

The Goshawk, well known in the wooded districts in Andalucia under the same name as the Sparrowhawk, is considered "muy valiente," being said to carry off partridges when they fall to the gun: this I know from my own experience. I can but consider it rare, having only met with the nest once, on the 15th of May, 1871, when I shot the female bird as she flew off the nest, which was a mass of sticks on an alder tree, about fifteen feet from the ground or, rather, mud, in the thickest part of the Soto Gordo, in the corkwood. The nest was evidently not a new one, and seemed to be an old nest of some eagle repaired by the Goshawks. It contained three eggs on the point of hatching, stained yellow all over with dirt, so as to resemble the eggs of a Grebe which had been sat on some time. On washing one of these eggs, however, the bluish ground-colour appeared.

I saw at Tangier several eggs stained in the same manner, marked as Goshawk's; and until I took their fac-similes myself I did not believe them to be genuine.

Lord Lilford took a nest of the Goshawk, with three eggs (which appears to be the usual number), in the Coto Doñana, in April or early in May.

10. ACCIPITER NISUS, Linn. The Sparrowhawk.

Moorish. Bou-umeira takouk (Cuckoo-Hawk). *Spanish.* Gavilan, Milano jaspeado (Marbled Kite).

" Is resident in the vicinity of Tangier, and common during passage in small flights, which pass to Europe during February, March, and April, returning in August and September."—*Favier.*

The Sparrowhawk is resident in wooded districts near

Gibraltar, though not in any great abundance, being most frequent in winter and during migration. I noticed it passing the Straits on the 28th of March, and have dates of nests obtained on the 13th of May, 10th of May, and 17th of May in different years, the first two nests containing fresh eggs; all were in tall trees, in the cork-wood.

11. BUTEO VULGARIS, Bechst. Common Buzzard.

Moorish. Kesir Eknah (*Favier*). *Spanish.* Arpella.

According to Favier, the Common Buzzard is seen in flights on passage in March and April, like *Milvus korschun*. I have seen them myself crossing the Straits on the 11th, 15th, and 24th of March.

On the Spanish side they are very abundant from November to the end of February. I never detected any remaining to breed near Gibraltar; but from a nest in a pine tree, containing two eggs, I shot one on the 29th of April near Seville.

In the cork-wood of Almoraima there are certain high trees which are the favourite resting-places of Buzzards. These trees are always chosen to command a good look-out, and are used winter after winter in succession; if one bird is shot, another takes its place. They are too lazy to annoy the sportsman; so, except now and then killing one for identification's sake, I never molested them. I once observed one feeding on the carcass of a donkey, in company with some Griffon Vultures.

12. BUTEO DESERTORUM, Daud. Rufous Buzzard.

Moorish. Khabas (Robber).

" Resident near Tangier, and found in considerable numbers on all sides. Their food consists of rats, mice, snakes, frogs, large insects, leverets, rabbits, and chickens. They nest on rocks, laying two eggs (in March and April) of a white or greenish-white colour, spotted with yellowish or reddish brown; sometimes these spots completely cover the thick end of the egg. The males sit in their turn. The irides are yellow; the third and fourth quill-feathers, equal in length, are the longest in the wing. Twenty-four eggs of this Buzzard have passed through my hands."—*Favier*.

This red-coloured Buzzard is, as above stated, common in Morocco. On the 26th of April, 1871, we found a nest on the top of a very tall olive-tree in a santo or burial-ground in Garbia, shooting both the old birds, one off the nest, which was like a Kite's and was lined with fresh olive-twigs and rags. It contained two eggs on the point of hatching; they were of a white colour, thinly marked all over with very small, short, reddish-black lines, and were more rounded than average eggs of either of the Kites, though I have seen eggs of both *Milvus ictinus* and *M. korschun* very like them.

In this santo, perhaps two acres in extent, were some of the tallest olive-trees I have ever seen, on which were, besides the Buzzard's nest, one of the Common Kite, with young, two of the Black Kite; and in a bramble-brake at the edge was a nest of Marsh-Harriers, with young. The day before, we took Black-Kite's eggs quite fresh, which shows the relative time of nesting of the above-named species.

I always saw this Buzzard in wooded districts, like our Common Buzzard, generally sitting on the bough of some dead tree; and this makes me wonder that Favier did not mention it as nesting on trees.

On the Spanish side of the Straits I never met with it; nor have I seen a specimen which could be referred to this species. It is slightly smaller in size, and easily recognized within a hundred yards or so, from the reddish colour. The immature birds would be harder to distinguish without handling them. Irides fine orange-yellow.

13. GYPAËTUS BARBATUS, Linn. Lämmergeyer.

Spanish. Quebranthuesos.

The Lämmergeyer is sparingly met with in the sierras of Andalucia, and, though I did not observe it about Ape's Hill, is no doubt found in the mountains on the African side of the Straits. The hill-district close to Algeciraz is the nearest place to Gibraltar that I have seen it; and a pair frequented the Sierra de Plata in March last; but we failed to discover the nest, though it would then have been too late to take the eggs, as the Lämmergeyer breeds very early, laying in January two eggs on high rocks. Generally seen alone. The cuneiform or wedge-shaped tail in relief against the sky will serve to identify it when on the wing. It has more of the habits of the *Neophron* than of the true Vultures.

14. AQUILA CHRYSAETOS, Linn.   Golden Eagle.

Moorish. Ogab.

"Is found on passage near Tangier, passing north in January and February, returning in July and August. Some remain to nest on rocks in March and April."—*Favier.*

I never met with the Golden Eagle in Andalucia, though I have seen one said to have been shot many years ago near Gibraltar. They occur in the mountains near Granada, and probably in others of the high sierras. I found in April a nest of an Eagle, apparently of this species, on a very high cliff near Jebel Moosa, opposite Gibraltar; but being unable to obtain the bird, we left the nest untouched.

15. AQUILA ADALBERTI, Brehm.   White-shouldered Eagle.

Spanish. Aguila real.

The White-shouldered Eagle is stated by Favier to be rare near Tangier. He calls it *Aquila imperialis,* and gives a local name ("Larnaj") describing an adult bird. I have examined specimens in immature plumage from there, and seen what I considered to be this Eagle on the wing.

In wooded districts in the west of Andalucia it is universally distributed, being most abundant in the Cotos towards Seville and about Cordova, not unfrequently occurring near Gibraltar. A tree-nesting Eagle; the old bird sits very close, but not more so than some other Raptores. Three eggs is the usual complement; and these are generally laid during the last fortnight in March, being usually white; they are sometimes much spotted with reddish brown, and vary much in shape and size. The nest is sometimes lined with horse-dung as well as green twigs. In the winter they mostly roost close to their nesting-places.

The adult birds are easily recognized on the wing from their dark appearance. The immature are less easy to distinguish; for a long time some of them were thought to be specimens of *A. rapax*; and I remember being considered a heretic in ornithological matters for saying they were young Imperial Eagles, at that time the difference between *A. imperialis* and *A. adalberti* not being known: my opinion, however, has since been proved correct.

The adults of this Eagle are very dark brown on the body and wings, except for the white patches on the latter, whence its name. The immature birds are at first of a uniform reddish brown, which becomes gradually lighter. They take, in captivity, three years to show any white in the wing. I am not aware of any instance of their breeding in the immature plumage.

16. AQUILA RAPAX, Temm. Tawny Eagle.

This Eagle is included in Favier's list; but as it is very doubtful if he obtained it, I omit his notes, and merely state that I never saw one on either side of the Straits.

17. AQUILA MACULATA, Gmel. Spotted Eagle.

The Spotted Eagle does not appear to have been obtained by Favier in Morocco. The only two specimens which I have seen from Andalucia were both from near Seville—one in spotted plumage (the same mentioned by Lord Lilford and Mr. Saunders in 'The Ibis'), and an adult male which I got, killed on the 12th of November, 1870, and now in Lord Lilford's collection.

18. NISAETUS FASCIATUS (V.). Bonelli's Eagle.

Moorish. Teir Thum (Garlic-bird).—*Favier.*

"This, the most common Eagle near Tangier, is resident there, though some migrate north in February and return in July. They are seen alone or in pairs hunting over a wide extent, feeding chiefly on hares and rabbits; they nest on rocks and high trees, laying in March one or two eggs, never more, of a rather round shape, rough and white in colour, with sometimes green and bluish stains. On the 29th April, 1867,

I took a nest containing one young female, which was able to fly on the first of July and was very savage.

"They are so voracious and plucky that I have known two instances in which they allowed themselves to be caught rather than give up their prey: one was taken by a Moor throwing his burnouse over the Eagle, which had struck down a tame pigeon; the other driving a fowl into some brambles, was caught before it would quit its prey."—*Favier*.

Bonelli's Eagle is found generally distributed as a resident in most of the mountain-ranges of Andalucia. I know of the sites of several nests, but, not wishing to make them public for the benefit of dealers, refrain from mentioning the exact localities, merely observing that they do not appear ever to breed together in the same range of cliffs, each pair holding their own district. One pair nest annually at Gibraltar, at the "back of the rock," to the south of the signal-station; there are never more than one pair, though there are four situations where there are nests, one of which has not been used for several years. These nests are built of sticks, and are placed on small ledges of the steep rock, with one exception well open to observation from the signal-station, where I used to spend many an hour watching the old birds and their habits. For some years they used two of the nests alternately, year about. The sergeant in charge of the signal-station, and the signal-men, one of whom had been there eight years, all agreed that they never knew two nests in one season, or saw more than one pair of old birds. Lord Lilford asked me to try and obtain the eggs for him; so in 1870 I made arrangements, by aid of the "almighty dollar," with some men who had been goatherds at Catalan Bay, to endeavour to secure the prize. They laid ropes down from the top to a bush-covered ledge, which was about two hundred feet above the nest; thence one man lowered himself; but unfortunately the nest was so overhung that, though he could nearly touch the eggs, he could not take them, so was obliged to reascend unsuccessful. The next day we arranged with improved gear to renew the attempt; but an officious official kindly reported me to the authorities as disobeying an ancient garrison order which prohibits animals and birds on the rock from being destroyed; so I had to eat "humble pie" and give the affair up as a bad business. The following notes as to the time of nesting may be interesting. Sergeant Munro, of the Royal Artillery, in charge of the signal-station, assisted me with two or three of the dates during my absence from the Rock.

In 1869, the eagles nested on the lower site, about 300 feet from the base of the Rock, which here ends on the steep sand slope south of the village of Catalan Bay.

In 1870 they used the upper nest, and two eggs were laid; the birds were sitting on the 20th of February; only one was hatched.

In 1871 the nest of 1869 was repaired, the birds beginning to renew it about Christmas 1870; two eggs were laid by the 6th of February, both of which proved fertile.

In 1872 the upper nest, that of 1870, was the favoured one: the repairs began on the 20th of December, 1871; the first of the two eggs laid was deposited on the 5th of February. On the 16th of March, both were hatched, making forty days occupied in incubation. Both birds sometimes sit at the same time; but usually they relieve one another. They continually turn the eggs over with their bills; and sometimes, when taken, the eggs bear marks of this in the shape of scratches. The upper part of these nests was always entirely rebuilt with fresh green olive-boughs, lined with smaller twigs of the same. Some of the boughs accidentally dropped I picked up at the foot of the Rock, gnawed through as if by rats. It must have cost the Eagles much time and trouble to procure them, as olive is very hard and tough wood.

In 1873 I was not at Gibraltar; but on my return in 1874, on the 24th of February, I found that they had built in a fresh situation near the other sites, and that two unspotted bluish white eggs, rather smaller than the usual type, had been taken the day previously by the aid of the same men whom I had employed in 1870. This nest was hid from view of the signal-station by a projection of the rock, and was easily obtained, the cliff there being less than half the height of that where the nest of 1870 is placed. In company with the officers who obtained these eggs, we took another nest of

Bonelli's Eagle at some distance from Gibraltar. It was on some rocks where the previous spring they had had the good fortune to take two eggs. We found the nest built in a different situation, easily obtained by the aid of a rope, and very neatly built and lined with twigs and leaves of the cork-tree; it contained two splendid eggs, beautifully marked with red streaks and spots, similar to those taken in 1873, and doubtless laid by the same bird. I was informed that the latter nest was lined with leaves of the asphodel, and that the spoilers literally walked into the nest. I saw the situation myself; and it was certainly the easiest to reach that I know of, as they usually build on the face of steep cliffs.

A nest which I found in 1874 contained only one egg, which was addled; but curiously enough the bird was sitting hard on this rotten egg, and I succeeded in shooting the female. This nest was in a hole, and only about 50 feet from the base of the steep cliff in which it was placed, and was lined with twigs and leaves of butcher's broom (*Ruscus aculeatus*). Not having enough rope to lower to the bottom of the rock, we had much trouble in getting this egg; however, we sent for more rope and lowered it down from above, tying a sack full of stones to the end to prevent it lodging in the rock; but after securing the object of our labours from below, we discovered that the rope, of which there was over 400 feet, had become fixed in the rock about halfway up, and no power would move it. The idiotic Spaniard whom we had left at the top, when he found that he could not pull it up, flung it down without tying a stone to the end; so it caught in several places; and by way of finishing he came down to where we were sitting, and, after pulling violently at the lower end, suddenly let go, when of course the rope flew up and lodged in the rock out of reach; so we had to leave it dangling about the cliff as a memorial or, rather, as a Spaniard remarked, "*un señal de los locos Ingleses.*"

The usual number of eggs of Bonelli's Eagle is two, and but rarely one; the colour is generally white, and I have only seen a few marked with red and buff spots and streaks.

At Gibraltar, Bonelli's Eagle may be often seen suspended,

as it were, in the air, head to wind, apparently immovable, like an artificial kite, for sometimes nearly two minutes. At this time, when watched through a glass, no movement of the wings can be noticed beyond an almost imperceptible quivering; but the legs and feet are continually shifted as if used to balance the bird. When not breeding, they hunt together, one high above the other, suddenly stooping down on some luckless rabbit or else gliding off to take up a fresh aerial station whence to watch for their prey, which seems to be always taken on the ground. They feed chiefly on rabbits, but have taken poultry away from the signal-station; and Sergeant Munro informs me that once one of the Eagles struck at and seized his cat, but let it go after cutting its back open and drawing blood.

Once at Gibraltar, in February, I watched two Ravens for a long time bullying one of these Eagles, which now and then made a futile dash at his tormentors, but at last turned tail, leaving the Ravens masters of the situation. On another occasion, in the same month, I saw a Bonelli's Eagle flying about not far from the Osprey's nest, when down swooped an Osprey, like a stone, striking the Eagle on the back and knocking out a lot of feathers. Shrieking out, they were bound together for a few seconds, and then separated, neither apparently the worse for the encounter, and each flying off towards their respective eyries. They were so close as to be within easy shot when (to use a Yankeeism) they "collided." A young bird about a month old was bought from a Moor at Tangier, and sent to me on the 18th of April; but it was so wild and savage that I thought it would kill itself. However, I succeeded in bringing it to England for the benefit of the Zoological Gardens.

The fully adult birds have a white patch on the back between the wings; and when viewed from above, this mark is very apparent and will at once identify the species; when below them the white appearance of the underparts and their very powerful gliding flight distinguish them. To a novice they mostly resemble the Osprey when on the wing; but the latter has a more flapping flight and shows its white head. Had the writer of a note to the 'Field' about the nesting of the Osprey at Gibraltar in 1868 or 1869 known this, he would not have credited the Ospreys with the nest of Bonelli's Eagle. The nest of the former, I may add, is far out of view of the signal-station.

The immature birds of this species are, in their first plumage, of a uniform rich deep brown colour, which becomes lighter with age, as with the young of *Aquila adalberti*.

The tarsus, feathered to the feet, is very long for the size of the bird, the thigh being still longer in proportion.

The cere and feet are pale yellow; the bill yellow, with a bluish-black tip; the irides golden yellow. A freshly killed female measured 27 inches in length, the wings from tip to tip across being exactly 5 feet, wing from carpus to tip 25 inches, tarsus $4\frac{1}{2}$ inches. I never heard any good distinctive local name for Bonelli's Eagle; perhaps the best is "el aguila de las rocas."

19. NISAETUS PENNATUS (Gm.). Booted Eagle.
Moorish. Ta-ferma (*Favier*).

"This Eagle is migratory, crossing to Europe in March and April, returning in September; some remain to breed in the vicinity of Tangier to go south for the winter with the return migration. They nest on high trees in April and May, laying from one to three white eggs, often much stained and with a rough surface. It is abundant when on passage."—*Favier*.

On the Spanish side this, the smallest of the European Eagles, is, about Gibraltar, entirely migratory. I noticed many crossing on the 24th of March. It frequents wooded districts, and is the most plentiful of the birds of prey in the cork-wood during the summer, when their wailing cry may be heard all day long. The nests there that I saw were, without exception, on oak trees, sometimes completely hidden in ivy. In the Cotos near Seville they generally build in pine trees. The same nest is used year after year; if the old birds be shot, next season another pair take possession to repair and reline it with fresh green twigs of the oak. Two is the usual number of eggs; I have known three, but frequently only one; their general colour is pale bluish-white, sometimes stained or spotted with faint buff marks. The earliest taken was on the 12th of April; but about ten days later is the best time to get them. This Eagle when put off the nest, instead of flying straight away, stoops down till it nearly touches the ground, and then flies away gradually rising.

Their principal food, judging from the examination of nests and the crops of specimens, appears to be young rabbits. They are easily recognized by their small size when on the wing, and by the light colour of the underparts. A local name which I have heard for them is "Bacallao," from the fancied but far-fetched resemblance in colour and shape which they are supposed to have when flying overhead to that staple article of Spanish diet, a split dried salt codfish; but I may as well mention that I cannot help thinking this name was fabricated for my special benefit.

The young birds generally are of a uniform dark reddish-brown colour; but this is not always the case.

The tarsi are feathered to the feet; the entire length varies from about 17 to 20 inches, depending upon sex.

20. CIRCAETUS GALLICUS, Gm. Short-toed Eagle.

Moorish. Tair el hisani (the Stallion-bird). *Spanish*. Culebrera (the snake-eater), Aguila parda.

"Migratory. Some remain to nest near Tangier, building on very tall trees or rocks, laying in April or May one egg, very round in shape, though slightly smaller at one end, of a white colour, sometimes marked with rusty spots. The males sit in their turn; the young do not fly till September. Those which pass over to Europe cross in March and April to return in October. Although not uncommon in the vicinity of Tangier, it is more so than Bonelli's Eagle. They will sometimes allow themselves to be killed on the nest rather than desert their young. Sixteen eggs have passed through my hands."—*Favier*.

The Short-toed—or, rather, it might be better named the Snake-Eagle, is common both in Morocco and Andalucia, frequenting wooded districts and the valleys of the Sierras, being by far the most abundant Eagle near Gibraltar, except the Booted Eagle, *Nisaetus pennatus*. As far as I could observe, they are migratory, as I never saw one in the winter months, the absence of their chief food (snakes and lizards) at that season being quite sufficient to account for their departure, as the temperature at that season, even in sunny Andalucia, is quite low enough to cause these reptiles to hibernate.

This Eagle breeds about the middle of April; all nests I have seen were in cork, oak, or pine trees, and consisted of a large mass of sticks, generally lined with fresh leaves and twigs of the cork tree. I found one exception to this among the ruins of the ancient city of Lixus near el Arish, or Larache, in Morocco, the nest being built in a thick mastick or lentiscus bush, the base of the nest actually touching the ground on the hill-side. In this instance there was no want of trees in the neighbourhood to account for the nest being placed in such an unusual situation.

On the 24th of April I shot the hen bird as she flew out of the bush. Had she remained quiet I probably should not have found her nest, which contained the usual single large white egg slightly incubated.

I never knew the Short-toed Eagle to nest in rocks, as Favier states, though I have often seen them perched on crags and large stones; but it is now well known that no absolute rule can be laid down as to the breeding of many species of the Diurnal Raptores exclusively on rocks or trees; they simply accommodate themselves to the country, even nesting on the ground if trees, rocks, or ruins are not available.

The cere, legs, and feet of the Short-toed Eagle are very pale greyish-yellow; irides yellow; inside of mouth pale blue. The long tarsus, bare of feathers, will alone serve to distinguish it from any other bird of prey of its size to be met with near Gibraltar.

21. MILVUS ICTINUS, Sav. Common Kite.

Moorish. Siwâna. *Spanish.* Milano real.

" Found in the vicinity of Tangier in much smaller numbers than the next species, being seen on passage only in pairs;

the birds which remain to nest appear to be those which are the first to go south; the remainder cross to Europe in March, returning in October; a few, however, stay throughout the winter. The eggs, two or three in number, are very similar to those of *M. korschun*, but always larger."—*Favier.*

The Common Kite is resident and to be seen almost everywhere on the Spanish side of the Straits—though in the immediate vicinity of Gibraltar they seldom occur except on passage, and are as common in winter as at any other season. They particularly affect districts where there are many pine trees, on which, in company with *M. korschun*, they nest, but from a month to at least a fortnight earlier, and never in such numbers as that Kite. The Common Kite is easily distinguished from the next species, when on the wing, by its light colour and much more forked tail; when flying overhead by the wings, which, underneath, are *light-coloured*, with one dark patch on each; in *M. korschun* the underparts of the wing are dark.

22. MILVUS KORSCHUN (Gm.). Black Kite.

Moorish. Siwâna. *Spanish.* Milano negro.

"Seen near Tangier in immense flights, which pass over to Europe in February and March, to return in August and September. Many remain to breed, awaiting the return migration from Europe, when they all disappear for the winter."—*Favier.*

Though a Spanish name is given above, very few Spaniards distinguish the difference between the common and the Black Kite; "black," however, is a misnomer, as the primaries are the only part of the plumage which is of that colour. The name *migrans*, by which this Kite is generally known, is most appropriate, as they are entirely migratory—the earliest day on which I observed them crossing the Straits being the 5th of March, then in great numbers, other days on which large flights passed being the 26th, 27th, and 28th of that month, some on the 23rd, one on the 29th of April, and six or seven on the 5th of May. The latest date of the return migration was the 9th of October. It is more abundant in the vicinity of Seville and where there are pine-woods; and very few pairs remain to breed about Gibraltar. Both in Morocco and in Andalucia they nest, often in colonies, about the end of April; and on the 24th of that month I took two nests near Larache, each containing the usual number of two eggs, both lots quite fresh. The nests, built of sticks and placed in tall trees like those of the common Kite, are lined with rags, paper, bits of rope, and such like rubbish.

The eggs are subject to great variation both in shape and colour; sometimes they are almost white, without any spots; others are richly marked all over with reddish brown; some only so marked at the ends, generally at the large one, though now and then at the smaller end.

23. PERNIS APIVORUS. The Honey-Buzzard.

Moorish. Khabas el graïn (*Favier*). *Spanish.* Aguila de Moros.

"Only observed near Tangier during passage, migrating north during April and May, returning in August and September. The autumnal migration is not in such great flights as the vernal one, the greatest number seen in autumn being from twelve to fourteen, usually six or eight, while in spring flights of many more than a hundred may be seen crossing the Straits in a body. Their plumage is so variable, it is almost impossible to find two exactly alike."—*Favier*.

The Honey-Buzzard, as above stated, is to be seen in swarms during the spring migration, which extends over some twenty days, being at its climax about the 8th of May. The latest flight I saw was on the 15th of that month. When they have once passed the water the passage is usually made in a gyrating flight of eccentric circles, sometimes very high and as often within shot of the ground. They seem, when thus circling onwards, as if about to alight; but I never saw them do so; nor have I ever seen them except at the period of migration. Lord Lilford observed large flocks passing south in September.

24. ELANUS CÆRULEUS, Desfont. Black-shouldered Kite.

Moorish. Aisha hemika (*Favier*).

"Scarce in the vicinity of Tangier, being seldom seen—and then in very limited numbers, in February and March and again during September and October. They are more common near Larache, where some are found breeding in April. They live on birds and small mammals, and are very voracious. Their cry is a sort of whistle."—*Favier*.

On the African side of the Straits I found the Black-winged Kite, as it is familiarly termed, common near Tetuan in April, as well as about Cape Negro; around Tangier at that time I only saw two. They nest on trees, and (as in other countries in which I have seen them) keep to slightly wooded places, not frequenting open ground.

On the Spanish side this Hawk is very rare. I never obtained one; but Lord Lilford records a specimen (Ibis, 1865, p. 177) as occurring near Seville. Easily recognized on the wing by its greyish-white colour. It has a peculiar habit of hovering at about thirty yards from the ground, with the wings forming a sort of V or acute angle with the body, never bringing them level with one another until it flies off to take up a fresh position. They are rather wary when thus engaged in hunting for their prey.

The general colour of this Hawk is grey above, white below, with black shoulders and crimson irides.

25. FALCO COMMUNIS, Gm. Peregrine Falcon.

Moorish. Teir el hor. *Spanish.* Alcon.

"Is not uncommon near Tangier, where some remain to breed; the remainder are migratory, going to Europe in February and March, returning in November and December. They nest from March to May on rocks and on trees, laying four eggs, eleven of which have passed through my hands for sale."—*Favier*.

The Peregrine Falcon is most abundant in Andalucia in winter; but some few are resident—a pair nesting at Gibraltar near O'Hara's Tower, and occasionally coming into the town to carry off tame pigeons. I think this pair belongs to the small race of Peregrine which inhabits the coast of the Mediterranean; but they certainly are not the Barbary Falcon. The usual average-sized Peregrine, however, nests on rocks on both sides of the Straits about the 21st of March, laying from three to five eggs. One breeding-place is about four miles from Tangier, in the middle of a colony of Rock-Doves (*C. livia*), with whom, as is usual in such cases, they live on apparently amicable terms. In this range of cliffs Bonelli's Eagle, the Osprey, and a few Lesser Kestrels also nest.

26. FALCO BARBARUS, Linn. Barbary Falcon.

The true Barbary Falcon undoubtedly occurs near Tangier, as I obtained one freshly killed specimen from Oleese, shot in the neighbourhood, and know of another skin in Mr. Dresser's collection; but I never myself shot one on either side of the Straits.

This species may be described as a miniature Peregrine with a rufous nape.

27. FALCO FELDEGGII, Schl. The Lanner.

Favier has, in his MS., under the head of *Falco barbarus*, evidently described the Lanner, as his measurements are larger than those of *F. peregrinus*, instead of smaller; and all the specimens of the Lanner which I have seen from Tangier, with one or two exceptions, were labelled "*barbarus*." Favier adds:—"This species, which the Moors confound with the Peregrine, is resident and as common as that species around Tangier."

On the 1st of May, 1872, I obtained a female Lanner and three eggs. The nest was on the rocks near the above town. Two of the eggs were slightly sat on; the third, much lighter in colour, was addled, which is often the case with eggs faintly marked or differing from the usual colouring.

On the Spanish side of the Straits I did not succeed in obtaining the Lanner; but it has lately been found nesting near Seville on pine trees, close to the Coto del Rey. In one instance, an old nest was used, from which three years previously I had shot a Buzzard (*B. vulgaris*) and taken two eggs. The last nest found contained eggs at the end of March.

The Lanner is distinguishable by the top of the head and nape being rufous in the adult and almost white in the young, though I have seen young Peregrines marked in the same manner.

The entire length is about 18 inches.

28. FALCO ELEONORÆ. Eleonora Falcon.

I have never met with this species on either side of the Straits; and there is no authentic record of a specimen having been obtained in Andalucia. Gilbert White's brother, the Rev. John White, writing from Gibraltar, mentions the Hobby as nesting at the "back of the Rock" more than a hundred years ago. If a Hobby did nest there, it could not well have been any species but the Eleonora Falcon, as the true Hobby (*Falco subbuteo*) is a tree-nesting bird. The Eleonora Falcon, however, occurs at Mogador, and nests there. It is distinguishable by its larger size from *F. subbuteo*, and differs also in its dark uniform plumage when adult.

29. FALCO SUBBUTEO. The Hobby.

According to Favier this little Falcon is seen near Tangier in pairs on passage only, "crossing to Europe in May, returning in autumn to winter further south."

Near Gibraltar the Hobby appears in the same manner; the earliest dates on which I noticed them were the 8th of April, 13th of April, and 20th of April, in three different years. I saw them near Seville very early in May; but there is not as yet any authentic record of their breeding there.

30. FALCO ÆSALON, Tunst. The Merlin.

Spanish. Esmerejon.

"Occurs during winter near Tangier, coming from Europe in September, returning north in March."—*Favier*.

The Merlin is not uncommon in open ground in Andalucia in December and January. The earliest I saw was on the 24th of November, the latest on the 7th of March. About

Casa Vieja they are most plentiful, and often to be seen chasing snipe and larks; they are, for the most part, adult blue-backed birds.

31. CERCHNEIS TINNUNCULUS (Linn.). Common Kestrel.
Moorish. Bou-umeira. *Spanish*. Cernicalo.

"Is both resident and migratory in Morocco. Those which migrate cross to Europe in February and March, returning in August and September. They nest by preference on old ruins and walls."—*Favier*.

It is needless to say much about this Kestrel, so well known at home. It is a resident both in Morocco and Andalucia, where it is very common, more so in autumn and spring, nesting in April on trees, rocks, and buildings.

This species may be distinguished from the Lesser Kestrel by its *black* claws and larger size. The adult male is spotted on the back, this part in the Lesser Kestrel being of a uniform cinnamon-rufous colour. Beyond Larache, in Morocco, the females seemed to me to be much darker than further north.

32. CERCHNEIS NAUMANNI (Fleisch.). Lesser Kestrel.
Moorish. Souif (*Favier*). *Spanish*. Primilla.

"Is nearly as abundant near Tangier as the Common Kestrel, passing to Europe in February and March, returning during August and September."—*Favier*.

The Lesser Kestrel is almost entirely migratory, though a few remain at Gibraltar during winter. Vast numbers nest there, chiefly on the steep face of rock on the North Front. These birds arrive about the 15th of February; but I saw a great flight passing as late as the 4th of April. Probably these were birds which would breed much further north. They nest on rocks and ruins, particularly on the old Moorish buildings and towers, of which there are so many in Andalucia. In some, as for instance at Las Alcantarillas, near Seville, they swarm like bees at a hive, as also at Seville; while, curiously enough, at Cadiz they are conspicuous by their absence.

As far as I am aware, the Lesser Kestrel never nests on

trees like the common species. At the Coto del Rey, on the 26th of April, I took a nest with four eggs out of a hole in a wall which I could reach from the ground. In the Crimea, I remember, they nested in holes of river-banks. On the 12th of May, near Marchena, we obtained sixty eggs out of an old tower, and might have taken as many more. Some of these eggs were hard sat on; and the old birds were caught on the nest, to be released after examination. These eggs varied very much, some being almost colourless, others half white, half red, piebald in appearance.

It is entirely an insectivorous bird.

33. CERCHNEIS VESPERTINA (Linn.). Western Red-footed Hobby.

Favier confounded this insectivorous Falcon with the Hobby, calling it a variety; he, however, gives a description which identifies it, and says "This variety is found near Tangier in April." It is certainly not common there, and is said only to appear when there are locusts, which they follow from the east. In 1871, on the 27th of April, I saw two near Tangier; shortly afterwards some were obtained by Olcese; and just at that time flights of locusts arrived. Curiously enough, in 1874, on the very same day in April I saw one close to Tangier, and the next morning saw quantities of locusts as we were crossing over to Gibraltar.

On the Spanish side of the Straits I never saw one, though it is recorded as having occurred near Seville; any way, it is a very rare bird so far west.

The adult male is of a dark lead-grey colour; thighs, vent, and under tail-coverts chestnut, with reddish-orange legs and feet.

The adult female has the head and the whole under-surface of the body rufous. The claws at all ages are nearly *white*.

Length from 11·5 to 12 inches.

34. PANDION HALIAETUS, Linn. The Osprey.

Moorish. Bou haut (Father of fish). *Spanish*. Aguila pescador.

"This bird is not uncommon near Tangier, living among the rocks on the coast, where they nest in March, laying two or three eggs; the young do not fly until July. The migrating birds arrive in October and November, returning north in March."—*Favier.*

The Osprey is most abundant in the Straits in winter. I saw a pair catching fish near Cape Negro, at Lake Esmir, in April; and a pair nest on the rocks westward of Tangier. Another pair regularly breed at Gibraltar, on the rocks a little to the north of " Monkeys' Cave." The Rev. John White noticed the nesting of the Osprey at Gibraltar about a century ago; and probably this is the same situation, and has been used ever since. I first knew of the eyry in May 1869, when there were young in the nest; these did not fly till the middle of July. In 1871 the nest was taken in the middle of March, and then contained three eggs; the old birds did not leave the vicinity, and bred again the next season, but in a different situation close to the old one. The first site of the nest was only to be seen from the Europa Advance Battery, where I spent many an hour watching the old birds with a telescope.

Being positive that only one pair of Ospreys breed at Gibraltar, and knowing the date of laying of that pair, I cannot account for the fact of seeing, on the 23rd of April, one take up from the surface of the sea and carry off a stick or splinter some three feet long; and on the 30th of March I also saw another carrying a stick. Could this be done in play? On the 17th of February I saw one of these Ospreys give a Gannet, which had ventured too near the nest, a great buffeting, knocking him about and chasing him for half a mile. The Isla de Palomas, a small patch of rock near the celebrated and dangerous Pearl Rock, is a favourite resting-place of these birds; and one is usually to be seen there at all seasons, perched on a small pinnacle.

A brother officer of mine killed an Osprey on the wing at Europa Mess-house with a pea-rifle. The bird was flying high up over the sea; but the very strong westerly wind blowing at the time caught and landed it among the men's huts; and it now (being well set up) remains a trophy of his skill with the rifle.

The cere, legs, and toes are blue.

Family STRIGIDÆ.

35. STRIX FLAMMEA. White or Barn-Owl.

Moorish. Youka (*Favier*). *Spanish.* Lechuza.

"This Owl, resident near Tangier, is nearly as abundant as the Little Owl, inhabiting ruins and holes in rocks, and nesting twice a year, between April and November. They lay from three to four eggs. The inhabitants of Tangier consider this bird the clairvoyant friend of the Devil. The Jews believe that their cry causes the death of young children; so, in order to prevent this, they pour a vessel of water out into the courtyard every time that they hear the cry of one of these Owls passing over their house. The Arabs believe even more than the Jews; for they think that they can cause all kinds of evil to old as well as to young; but their mode of action is even more simple than that of their antagonists the Jews, as they rest contented with cursing them whenever they hear their cry. Endeavouring to find out from the Mahometans what foundation there is for the evil reputation of this species, I was told this:—'When these birds cry, they are only cursing in their language; but their malediction is harmless unless they know the name of the individual to whom they wish evil, or unless they have the malignity to point out that person when passing him; as the Devil sleeps but little when there is evil work to be done, he would infallibly execute the command of his favourite if one did not, by cursing the Owl by name, thus guard against the power of that enemy who is sworn to do evil to all living beings.' Having learned the belief of the Mahometans relative to this Owl, it was more difficult to find out exactly that of the Jews, who when questioned by me knew not how to answer, except that the act of pouring out water in the middle of the courtyard is a custom of long standing in order to avert the evil which the Owl is capable of doing; that is to say, the water is poured out with a view of attracting the Evil Spirit's attention to an object which distracts him, and so hides from him the infant which the Owl in its wickedness wishes to show him."—*Favier*.

On the Andalucian side of the Straits the White Owl is common and resident, nesting at Gibraltar in the Moorish Castle.

I must here digress to say a few words in favour of this most useful of birds. Almost exclusively feeding on rats and mice, they deserve every encouragement and support that can be afforded them; but from being in all countries regarded with superstitious awe and dislike, they are more or less persecuted on that account; and in England, through the ignorance and stupidity of game-keepers, who fancy that they kill game (*i. e.* feathered game), they suffer most severely. This excuse is ridiculous; for the old birds they have not the power to kill, and young pheasants and partridges at the time the Owls are on the feed are safely being brooded by the parent bird.

Those who wish to encourage and increase Owls, and have not hollow trees or buildings where they nest, may always gratify their wishes by fixing an empty barrel (about an 18-gallon size) horizontally in the fork of any large tree, cutting a hole in one end large enough for the birds to enter; but the hoops of the cask should be screwed on, or it will soon fall to pieces. Not only the Barn-Owl, but the Tawny Owl (*Syrnium aluco*) also will use these barrels or "owl-tubs." The difficulty, however, is, to keep out the Jackdaws; but when once the Owls have established themselves, there is no fear of that intrusion.

In a barrel put up too near another in which was an Owl's nest, a pair of Stock Doves took possession and reared their young. This same tub afterwards had a hornet's nest in it.

36. SYRNIUM ALUCO, Linn.   Tawny Owl.

Moorish. Lŭ Lŭal, Bŭ-rŭ-rŭ.

"This species is the scarcest of the Owls near Tangier, being met with on passage, crossing to Europe in February, returning in November and December. Some remain to nest in April, laying two eggs, of which often only one is hatched. They live in large thick woods."—*Favier*.

Specimens of the Tawny Owl which I have seen from

Morocco were of the grey variety. I never met with or heard one in Andalucia, though no doubt, as it is in some of the Institutos, it does occur. The Arabic name Bŭ-rŭ-rŭ is delightfully suggestive of the cry.

37. CARINE NOCTUA, Retz. Little Owl.

Moorish. Mouka, Mouéka, Bouma. *Spanish.* Mochuelo.

"Is the commonest species of Owl near Tangier, being both resident and migratory. Those which migrate pass to Europe during March and April, returning in August and September. During passage they are met with in pairs or small flights; at all other times they are found singly or in pairs among large rocks and old buildings."—*Favier.*

The above was written by Favier under the head of *Athene glaux,* or, as he had it, "*Strix noctua meridionalis;*" but as that species has not yet been noticed in Andalucia, and as *C. noctua* is the Little Owl of Tangier, I have no hesitation in referring the above notes to *C. noctua*—not that it would have been much loss to have omitted them altogether, the only information of importance being that they migrate.

Near and at Gibraltar the Little Owl is common and resident, nesting, about the end of April, in holes of trees as well as in rocks.

Length 8·2 to 9 inches; tarsus 1·1.

38. CARINE GLAUX, Sav. Southern Little Owl.

This species, which is, in my opinion, only a light-coloured race of *C. noctua,* does not appear to be met with in the immediate vicinity of Tangier, the only specimens I have seen having been obtained three or four day's journey on the way to Fez. The Little Owl of Tangier is undoubtedly *C. noctua,* as well as that of Andalucia, where as yet *C. glaux* has not been met with.

39. SCOPS GIU, Gm. Scops Owl.

Moorish. Maroof. *Spanish.* Corneja.

"Occurs near Tangier on passage, crossing to Europe in March, returning to winter further south in September and

October. Many pass the breeding-season in Morocco."—*Favier*.

The Scops Owl is very plentiful, both in Morocco and Andalucia, but is almost entirely migratory. I was much surprised to hear one on the 13th of January, 1872, near the Coto del Rey, and another nearer Seville on the 15th; but from what I afterwards heard at Seville, there is no doubt a few sometimes remain there during the winter; I never heard them during that season at Gibraltar, the earliest date of the vernal migration noticed being the 4th of March, the first nest being on the 4th of May. This Owl always nests in holes of trees. I do not know of any instance of its nesting in rocks or ruins, like *Carine noctua*, which nests by preference in those places.

Abundant in the cork-wood; the nest is easily discovered by going round and hammering at the old cork-trees with a stick, when, if a Scops Owl flies out, ten to one there is a nest. They are strictly arboreal; and their monotonous single note may be frequently heard repeated at regular intervals by day as well as by night. They frequent trees in the midst of towns, and may often be heard in the trees which fringe the Delicias, the drive and Rotten Row of Seville.

They chiefly feed on coleoptera, and, I believe, are almost entirely insectivorous.

The irides, like those of most of the Tufted Owls, are yellow.

40. BUBO IGNAVUS, Forster. The Eagle-Owl.

Spanish. Bujo real.

This Owl is not included in Favier's notes on the birds of Tangier, though it is, no doubt, found in the mountainous districts of Morocco; indeed I heard of a large Owl about Tetuan, but could not obtain a specimen.

It is resident in all rocky localities in Andalucia; and some frequent the " Rock," probably nesting in some of the numerous inaccessible caverns of the east side. One was caught in 1869 in a magazine near the Rock gun; having gone down into the narrow space between the outer and the main wall of the magazine, it was unable to rise. I had this bird alive for some time, and ultimately sent it to Lord Lilford, in whose possession it paired with another from Norway. I also had three young from a nest near Castellar, about eighteen miles from Gibraltar. When I had these Owls, the wild ones used to come at night close to the cage and answer the call of those that were shut up. Its loud, melancholy, human-sounding note is sometimes to be heard all night long up the Rock, and is usually supposed to be the cry of the apes.

They breed very early: judging from the size of the young which I obtained, they would lay about the end of January; and such is, I was informed, the case. I never could succeed in discovering the nest. I know of several reputed nesting-places, but on examining them found nothing but bones of rabbits, rats, partridges, and small birds, never even seeing one of the Owls, though the charcoal-burners (or *carboneros*) assured me that they had taken the young from these places. One man, however, said that these Owls bring the young from their nests to these caves. The Rev. John White mentions the Eagle Owl as occurring at Gibraltar during his residence there about 1776.

41. ASIO OTUS, Flem. Long-eared Owl.

Spanish. Carabo.

Not obtained near Gibraltar, this tree-haunting Owl is more common towards Cordova and Granada. I only met with it once in winter, in the Coto del Rey.

42. ASIO ACCIPITRINUS (Pall.). Short-eared Owl.

Moorish. El hama, Fav. *Spanish.* Carabo.

" This species occurs less abundantly than the Cape-Owl (*A. capensis*), being found on passage in small flights on open and wet ground. Some breed near Tangier; but the remainder cross to Europe in February and March, returning in November. This Owl interbreeds with the Cape-Owl, producing hybrids which only differ from that species in having the front of the facial disk, the throat and tarsi whitish, while the irides are *half* yellow. The Arab chasseurs confound the two species under the name of ' el hama;' but they are easily distinguished by the irides, which are yellow in the present species, and hazel in the Cape-Owl."—*Favier*.

The above story about the hybrids is difficult to believe, and is to my mind apocryphal; however, it is given for what it may be worth. I confess I am very sceptical as to the assertions made about the interbreeding of different species in an absolute state of nature, excepting only the Gallinæ; but most hybrids among these are produced under circumstances of acclimatization which can hardly be called a really wild state.

The Short-eared Owl may nest so far south, but as far as my observations go, is in Andalucia only a winter resident, and even then not very abundant. I should have omitted this story of Favier's but for its having appeared in print before.

43. ASIO CAPENSIS (Smith). Marsh-Owl.

Moorish. El hama.

" Is a common resident near Tangier, usually frequenting wet swampy ground, feeding chiefly on insects. Some pass over to Europe in March and April, returning in November and December. They nest on the ground in April or May, laying four, rarely five, round white eggs, sometimes marked with a few rusty spots. The young are are not always hatched at the same time, as in the same nest may be found young birds of different growths."—*Favier*.

My experiences of this Owl in Spain are very limited, and as follows:—In October 1868, on my first visit to Casa Vieja, when looking for snipe in one of the wettest parts of the Mill soto, two Owls rose at my feet, which I shot, winging one, which I carried home alive to take to Gibraltar, seeing at once from the bluish black colour of the irides that I had got an Owl which I did not know. Afterwards hunting about, only one more was seen, and killed. On the 10th of November following, during my second visit, I saw three more, two of which I winged and also carried off alive to Gibraltar, keeping them there for some time, till one got out and flew off as if nothing was the matter with it; so I sent the other at once to Lord Lilford, who had it alive till 1870. I met with no more till the 10th of November, 1870, when I shot one and picked up the remains of another. In October and November 1871 I repeatedly and carefully went over the same ground, but did not see any, while friends of mine there in August and September, whom I begged to look out for these Owls, did not come across one. All the eight birds above mentioned were found within a space of about a square mile; and, strange to say, I never saw any elsewhere. In December 1873, Lieutenant Reid, of the Royal Engineers, shot one when snipe-shooting in the same locality. I was there in March and May in 1874, and, though I hunted all the likely ground over, failed to meet with even one.

Order PICARIÆ.

Family CAPRIMULGIDÆ.

44. CAPRIMULGUS EUROPÆUS, Linn. Common Goat-sucker or Nightjar.

"Occurs near Tangier, but is less numerous than the Rufous-naped Goatsucker, some, however, remaining to nest. The others pass on across the Straits during May and June, returning from September to November to pass the winter further south."—*Favier*.

The Nightjar is found in Andalucia, as in Tangier, the earliest date of arrival noticed being the 5th of May.

45. CAPRIMULGUS RUFICOLLIS, Temm. Rufous-naped Goatsucker.

Moorish. Tref el royau (Favier). *Spanish*. Zumaya, Papa Vientos, Chota Cabras.

"This Goatsucker is very abundant near Tangier, arriving to cross the Straits in April and May, to return in October and November. Many remain to breed, nesting on the bare ground among scrubby brushwood, and laying two eggs, which are to be found from May to August."—*Favier*.

The Rufous-naped Goatsucker is extremely plentiful near

Gibraltar. I know one instance of its occurrence on the 16th of February near Malaga; the earliest date of arrival near Gibraltar was on the 15th of April, the latest date of departure the 5th of October. This species is easily known from the common Goatsucker by its larger size and by the light rufous markings at the nape of the neck; the eggs only differ in being slightly larger.

Family CYPSELIDÆ.

46. CYPSELUS APUS, Linn. Common Swift.

Moorish. Tair abila (*Favier*). *Spanish.* Avion.

"This Swift arrives at Tangier, *en route* for Europe, during March and April; vast numbers remain to nest here, and return south in September and October."—*Favier.*

The above notes equally apply to the common Swift in Andalucia, the earliest date of arrival noted being the 4th of March, the main body passing during the last fortnight in that month, some as late as the 24th of April. The majority leave by the end of August, some staying on into the middle of September, the last being seen on the 16th of October. The number in some towns, particularly Algeciraz, is perfectly marvellous, and the noise they make morning and evening quite annoying.

47. CYPSELUS PALLIDUS, Shelley. Mouse-coloured Swift.

"This Swift is found near Tangier on passage, crossing to Europe in April and May. Some remain to breed; but it is the least common of the family, being seen alone or in pairs in company with *C. apus*, which circumstance makes it difficult to distinguish them. I found a pair in July 1861, nesting in company with some House-Martins (*Chelidon urbica*); the nest was simply an old nest of that Martin, which the Swifts had appropriated, and contained two eggs of the usual *Cypselus* shape, their longitudinal circumference being 64–66 millimetres."—*Favier.*

Though Favier says they are difficult to distinguish from the common Swift, I cannot say so myself, but rather the contrary. They are occasionally seen near Gibraltar, and are said to arrive at Tangier somewhat earlier than the common Swift, though I could see no difference in the time of their arrival. Easily noticed on the wing by their light colour, they mix both with the common and White-bellied Swifts.

In May 1874, when near Vejer with two ornithologizing friends, we found this species to be more abundant than *C. apus*, while, curiously enough, at Algeciraz (where, as mentioned, there are countless swarms of common Swifts) I never could detect one single *C. pallidus*.

The general colour of *C. pallidus* is a uniform grey brown, except the throat, which is white, as are the tips of the feathers of the lower part of the breast, giving it a mottled appearance.

48. CYPSELUS MELBA, Linn. White-bellied Swift, Alpine Swift.

Moorish. El namera. *Spanish.* Avion.

" Found near Tangier on passage, crossing the Straits from March to May, returning from August to October. It is not so common as *C. apus*."—*Favier.*

The White-bellied Swift breeds sparingly at Gibraltar in the inaccessible crevices of the rocks on the Mediterranean side; they seem to arrive, if any thing, a little later than the common Swift. The earliest date of arrival noticed was the 24th of March : many were seen on the 29th. On the 4th of April, 1871, near San Roque, I noticed a flock of about two hundred passing in a northerly direction, with a gyrating flight, making a great noise, though they were very high up. A few common Swifts (their cry attracted my attention to them) were with the flock. On the 5th of November, 1871, at Casa Vieja, I saw six hawking about over the marshes for about an hour, when they disappeared in a westerly direction. I fired several shots at them, but ineffectually, as they were too high. An officer who was at Fez told me that he saw a great many Alpine Swifts there in large flights about the 23rd of February.

The sexes are alike in plumage, except that the female

is marked or striated with a faint black line on the centre of the feathers of the white breast. The brain of this, as in all Swifts that I have examined, is small for the size of the bird.

Family CORACIIDÆ.

49. CORACIAS GARRULA, Linn. The Roller.

Moorish. Sharrakrak. *Spanish.* Carlanco, Carraca.

"This bird is seen in numbers near Tangier on passage, migrating in pairs and crossing the Straits in April and May, returning in August to retire further south. Their food is all kinds of insects, even scorpions."—*Favier*.

The Roller breeds at Larache, nesting in holes of the walls of the ramparts at the end of April. I did not observe any elsewhere in Morocco, except about the ruins of "old Tangier."

In Andalucia they are also very local. I have seen one or two in May near Casa Vieja; but they are not common nearer to Gibraltar than the vicinity of Seville. Thence along the valley of the Guadalquivir to Cordova they abound. I never saw one about Gibraltar. They arrive during the latter end of March, leaving by September. Nesting in holes of trees, walls, and ruins, they lay (about the 14th of April) from four to six shining white eggs.

Family MEROPIDÆ.

50. MEROPS APIASTER. The Bee-eater.

Moorish. El Lecamoon. *Spanish.* Abejaruco.

"The Bee-eater is seen on passage near Tangier in great flights, which attract notice from their cry. They arrive and cross over to Europe during March and April, returning in August, many remaining to breed. They nest in May, the eggs varying in shape, being some oval, some oblong."—*Favier*.

This bird did not appear to me to be quite so common in Morocco at the end of April as on the Spanish side of the Straits, where during April, May, June, and July it is one of the most conspicuous birds in the country; at that season

Andalucia without Bee-eaters would be like London without Sparrows. Everywhere they are to be seen; and their single note, *teerrp*, heard continually repeated, magnifies their numbers in imagination. Occasionally they venture into the centre of towns when on passage, hovering round the orange-trees and flowers in some patio or garden. Crossing the Straits for the most part in the early part of the day, flight follows flight for hours in succession. When passing at Gibraltar they sometimes skim low down to settle for a moment on a bush or a tree, but generally go straight on, often almost out of sight; but their cry always betrays their presence in the air.

My dates of their first arrival noticed are:—the 7th of April, 1868; 4th of April, 1869; 1st of April, 1870; 29th of March, 1871; 26th of March, 1872; 28th of March, 1874. They were observed passing in great numbers from the 10th to the 14th of April in three consecutive years, the greatest quantity arriving on the 10th; so, in Spanish fashion, I christened that date "St. Bee-eater's day." The latest flight I ever saw going north was on the 7th of May.

Having remained at Gibraltar once only during July and August, I had but that opportunity of watching the return migration, which appeared during the last week in July and also on the 11th and 12th of August, the last being noticed on the 29th of that month, all with few exceptions being heard passing at night. The first arrivals, as is the case with all migrants, are those which remain to breed in the immediate neighbourhood. Commencing their labours of excavation almost immediately they arrive, the earliest eggs that I know of were taken on the 29th of April; but usually they do not lay till about the second week in May, often not so soon.

In some places they nest in large colonies; in others there are perhaps only two or three holes. When there are no river-banks or barrancos in which to bore holes, they tunnel down into the ground, where the soil is suitable, in a vertical direction, generally on some slightly elevated mound.

The shafts to these nests are not usually so long as those in banks of rivers, which sometimes reach to a distance of eight or nine feet in all; the end is enlarged into a round sort of chamber, on the bare soil of which the usual four or five shining white eggs are placed; after a little they become discoloured from the castings of the old birds, the nest being, as it were, lined with the wings and undigested parts of bees and wasps. Vast numbers of eggs and young must be annually destroyed by snakes and lizards: the latter are often seen sunning themselves at the entrance of a hole among a colony of Bee-eaters; and frequently have I avenged the birds by treating the yellow reptile to a charge of shot. The bills of Bee-eaters, after boring out their habitations, are sometimes worn away to less than half their usual length; but as newly arrived birds never have these stumpy bills, it is evident that they grow again to their original length. It has often been a source of wonder to me how they have the strength to make these long tunnels; the amount of exertion must be enormous; but when one considers the holes of the Sand-Martin, it is perhaps not so surprising after all.

During my stay at Gibraltar, Bee-eaters decreased very much in the neighbourhood, being continually shot on account of their bright plumage to put in ladies' hats. Owing to this sad fashion, I saw no less than seven hundred skins, all shot at Tangier in the spring of 1874, which were consigned by Olcese to some dealer in London. However, the enormous injury these birds do to the peasants who keep bees, fully merits any amount of punishment; but at the same time they destroy quantities of wasps. After being fired at once or twice, they become very wary and shy at the breeding-places; and the best way to shoot them is to hide near the *colmenares*, or groups of *corchos* or cork bee-hives, which in Spain are placed in rows sometimes to the number of seventy or eighty together; and it is no unusual thing to see as many Bee-eaters wheeling round and swooping down, even seizing the bees at the very entrance of their hives.

The reason of their early departure in August is to be accounted for by the simple fact that bees cease to work when there are no flowers; and by that time all vegetation is scorched up.

Family ALCEDINIDÆ.

51. ALCEDO ISPIDA, Linn. The Kingfisher.

Moorish. Kandil el behar (Light of the sea). *Spanish.* Martin pescador.

"This bird, only found from August to March, is not numerous near Tangier, but is more abundant near Rabat."—*Favier.*

The Kingfisher is common in winter and spring near Gibraltar, and is frequently seen among the rocks on the coast, and often at the "inundation" at the North Front. I have no record of its occurrence during the breeding-season—that is, not later than *the end* of April. The majority arrive in October, leaving in March.

Family UPUPIDÆ.

52. UPUPA EPOPS, Linn. The Hoopoe.

Moorish. Hudhud. *Spanish.* Abubilla, Gallo de Marzo, Cajonera, Cagajonera, Sabubilla.

"Seen in great quantities near Tangier on passage, crossing to Europe during February, March, and April, returning, to retire altogether for the winter, in August, September, and October. In some years the vernal migration is earlier, and they are seen at the end of January. They rarely remain to nest near Tangier. The females have a nearly white throat. The superstitious Jews and Mahometans both believe that the heart and feathers of the Hoopoe are charms against the machinations of evil spirits."—*Favier.*

Hoopoes seldom remain to nest in the vicinity of Gibraltar; but a few breed about Casa Vieja, and thence northwards, where there are trees; towards Moron and Seville their "hood, hood" may be frequently heard in spring and summer. They begin to lay about the 1st of May, in holes of trees.

My dates of their earliest arrival at Gibraltar are:—the 17th of February, 1870; 18th of February, 1871; 16th of February, 1872; but on the 11th of January in that year I saw a single Hoopoe in the Coto del Rey, but was unable to secure it. Lieutenant Reid also informed me of one appearing as early as the 16th of January in 1874. They mostly pass in March, whence their local name *Gallo de Marzo*, March-cock.

Family CUCULIDÆ.

53. CUCULUS CANORUS, L. The Common Cuckoo.

Moorish. Takouk, Oukouk. *Spanish.* Cucu.

"More abundant near Tangier than the Great Spotted Cuckoo; seen during passage, in pairs, which cross to Europe in April and May, and return in August to winter, probably, in the interior of Africa. Some remain during summer, awaiting the return of the autumnal migration."—*Favier.*

The Cuckoo is very plentiful near Gibraltar, especially in the Cork-wood and on all hill-sides wherever there are any trees. I saw a great many at the top of the mountains at the back of Algeciras at the end of May, but not beyond the line of trees. I first heard it on the 7th of April in 1868, on the 22nd of March in 1870, on the 31st of March in 1871, on the 29th of March in 1872, and on the 30th of the same month in 1874. They remain till the end of July.

A female shot in the second week in May had then two eggs remaining in the ovaries, nearly ready to lay.

54. COCCYSTES GLANDARIUS (L.). The Great Spotted Cuckoo.

Moorish. Teir el Kheber (Bird of news) (*Favier*). *Spanish.* Cuco real.

"Occurs near Tangier on passage, always in pairs, but not in any great numbers. They cross to Europe in January, February, and March, returning in June, July, August, and September. Their food is entirely caterpillars, both smooth and hairy."—*Favier.*

The Great Spotted Cuckoo arrives in Andalucia much earlier than the Common Cuckoo, *Cuculus canorus*; and though Favier states that they pass in January, the 25th of February and the 2nd of March are the earliest dates which I have for their arrival, and they mostly appear between the 7th and 28th of March. The latest I saw was on the 7th of August, in the Alameda at Gibraltar; but they are seldom noticed near there, and pass on to districts further north, where there are Magpies (*Pica rustica*), as they lay in the nests of the latter, and occasionally, it is stated, in those of the Spanish Magpie (*Cyanopica Cookii*). The egg can be easily distinguished by its elliptical form, those of the Magpie being pointed at one end. They vary a good deal in size and much in the markings, like those of the bird whose nest they use. It appears that, as far as we yet know, this Cuckoo always places its eggs in the nests of the *Corvidæ*. The majority of eggs I have seen, mostly obtained by Ruiz of Seville, came from the vicinity of Cordova; there are a good many in the Coto del Rey, where I had the satisfaction of shooting the first one of these birds I ever saw alive. The Rev. John White mentions this Cuckoo as having been killed at Gibraltar. A female killed on the 7th of March had the eggs so far developed as to show that the probable number of eggs she would have laid was four.

The irides are brown; there is no difference in the plumage of the sexes. The young have rufous secondaries and black heads, and are more handsome birds than the adults.

Family PICIDÆ.

55. PICUS MAJOR, L. The Great Spotted Woodpecker.

This Woodpecker is very local, and is always to be seen or heard among the old alder trees in the Soto gordo of the cork-wood of Almoraima. They extend all over that wood, up the valleys of the sierras, particularly along that one down which runs the river Palmones. They are common near Ojen, also abundant about Pulverilla on the road between Casa Vieja and Gibraltar. Further than this I never noticed them; nor did I ever see them in any country where oak and alder trees were absent. The local name is *Pito real*. They nest about the first of May in holes of decaying trees, and do not appear to be in the slightest degree migratory.

The adult male has a crimson occipital patch. The adult female has all the top of the head *black*, while the young are the more gaudily marked, having the top of the head entirely crimson. All have the forehead a dirty whitish brown, varying a great deal according to the weather, being mostly stained from the wet rotten wood in which they seek their food.

56. PICUS NUMIDICUS, Malh.   The Algerian Pied Woodpecker.

Moorish. Nakab.

"Resident and common in the vicinity of Tangier, being found only in large woods, where they nest in holes of trees, laying from five to six eggs, similar to those of *P. major.*"— *Favier.*

I did not find this bird " common" near Tangier; and as for the " large woods," there are none close to that town; about Tetuan this Woodpecker is plentiful, similar in habits to *P. major*. Favier states that they migrate across the Straits; but I should say this can hardly be the case. I have seen and shot many specimens of *P. major* in Andalucia, but never met with *P. numidicus*, although three or four of the Spanish birds had some few crimson feathers on the breast.

The difference between the species, as given in Sharpe and Dresser's ' Birds of Europe,' is that the African bird has at all ages a crimson pectoral band and a longer and more slender bill, while the young has a *black* forehead.

57. GECINUS SHARPII, Saunders, P. Z. S. 1872, p. 153. The Spanish Green Woodpecker.

Spanish. Pito real.

This Green Woodpecker, in habits, note, and manner of nesting, is exactly similar to the British *G. viridis*, and only differs from it in having the sides of the face dark grey instead of black, and in being brighter-coloured on the head and rump, the latter being much brighter, and in some individuals almost flame-coloured; but I have seen British-killed specimens of *G. viridis* quite as highly marked. The female, as in that species, has no crimson moustache. It is abundant in some localities near Seville, particularly in the Cotos and towards Cordova and Granada. I never met with it nearer to Gibraltar than the vicinity of Seville.

58. GECINUS VAILLANTII. The Algerian Green Woodpecker.

Moorish. Nakab el debak (The borer of the bark).

" Resident near Tangier, but not so common as *Picus numidicus*; like the latter avoiding the haunts of men and living in large woods. They nest in holes of trees in April and May, and lay from five to eight shining white eggs. The males assist in incubation."—*Favier.*

I found this Green Woodpecker to be common near Tetuan and in the province of Angera, especially among the short stunted trees which grow in the valleys about Jebel Moosa; nearer towards Tangier it is rare, the scarcity of trees accounting for its absence; in habits and in note it agrees with G. *viridis.*

The marks which distinguish it from that species are the same as those in G. *Sharpii*, in addition to which the moustachial stripe of the adult male, which is red in both G. *viridis* and G. *Sharpii*, is black in the present species; specimens from Tetuan agree in size with both the Spanish and Common Green Woodpecker. It is needless to add I have never seen this African form on the European side.

Irides white.

59. GECINUS CANUS, Gm. The Grey-headed Green Woodpecker.

I have never encountered this bird in Andalucia; but there is or was a specimen in an Instituto at Seville, said to have been obtained in the neighbourhood, and Lord Lilford records it from the vicinity of Madrid.

It is easily distinguished by the grey head and neck, the adult male only having a crimson patch on the fore part of the head.

60. YUNX TORQUILLA, L. The Wryneck.

Spanish. Torcecuello.

" Rather scarce and seen only in pairs near Tangier during passage, crossing the Straits in March and April, returning in August and September, but occasionally observed up to December."—*Favier.*

On the Spanish side of the Straits, I have seldom met with the Wryneck near Gibraltar, and only in March and September; probably their line of migration lies further to the east.

Order PASSERES.

Family TURDIDÆ.

61. TURDUS MUSICUS, Linn. The Common Thrush.

Spanish. Zorzal.

Favier's note applies to this bird on both sides of the Straits, and is as follows:—"The Song-Thrush is a winter resident in great numbers, being the most common of the Thrushes, arriving in large flocks in October and November, departing in March." They chiefly frequent the wild olive-trees, on the berries of which they feed. The first date of arrival noticed at Gibraltar was the 22nd of October; and the latest day on which I observed them was the 1st of April.

62. TURDUS VISCIVORUS, Linn. The Missel-Thrush.

Spanish. Charla (Chatterer).

"Found near Tangier, always singly and very sparingly in company with *T. musicus*, on passage. They arrive in November but do not stay near here, returning to recross the Straits in February."—*Favier*.

The Missel-Thrush cannot be said to be common near Gibraltar, being most so in winter. They are considered to arrive and depart with the Woodcocks; but they occasionally (as in 1870) remain to nest in the cork-wood, where I also saw a pair on the 4th of April 1871, and again in the end of April 1874.

63. TURDUS PILARIS, Linn. The Fieldfare.

Is not mentioned in Favier's list, and is therefore no doubt of very rare occurrence in Morocco. I obtained one specimen from there, a curiously plumaged young bird, figured and described in Sharpe and Dresser's 'Birds of Europe.' On the Spanish side of the Straits I never met with it.

64. TURDUS ILIACUS, Linn. The Redwing.

"This Thrush is very rare near Tangier. I have only met with two, between November and March—one in 1852, the other in 1864."—*Favier.*

The Redwing is rare in the vicinity of Gibraltar in winter, and I never recollect seeing it, though Mr. Saunders states that it was nearly as abundant as the common Thrush at Malaga in the winter of 1867-68.

65. TURDUS MERULA, Linn. The Blackbird.

Moorish. Tchau Tchau (*Favier*). *Spanish.* Mirlo.

"Resident near Tangier and very plentiful, nesting three times a year."—*Favier.*

I found a nest in Morocco built in a prickly-pear hedge. The Blackbird nests at Gibraltar, and is very common in Andalucia, more so in the winter months.

66. TURDUS TORQUATUS, Linn. The Ring-Ouzel.

Spanish. Chirlo.

"Is only met with in small flights on passage near Tangier, crossing to Europe in March and April, and returning in the autumn to pass the winter further south."—*Favier.*

I only observed the Ring-Ouzel near Gibraltar on passage in the spring, the earliest dates in each year being the 8th of April 1868, 20th of March 1870, 9th of April 1871, 12th of March 1872, 28th of March 1874.

67. PETROCOSSYPHUS CYANUS, Linn. The Blue Rock-Thrush.

Moorish. Tchau-tchau zerak. *Spanish.* Solitario.

Favier states that the Blue Thrush, which is as common in suitable localities in Morocco as in Andalucia, is migratory, passing north from February to May, and passing south from August to September. I never could detect any migration on the Spanish side, and consider it one of the very few birds which are stationary, not even shifting their ground—though, perhaps, in other countries circumstances may cause them to migrate.

Abundantly distributed on all rocky ground, even on sea-cliffs, and often seen on house-tops in those towns which lie in their districts, they are always to be found at Gibraltar in unvarying numbers, frequenting daily the same spots, and attracting considerable notice both from their agreeable song and conspicuous habits.

I here repeat a note I made about their nesting, which has already appeared in Sharpe and Dresser's account of this species:—" A pair nested in a hole outside the wall of my stable at Gibraltar in June 1869. Five eggs were laid, which were hatched about the 20th. The nest, composed of small dried bits of roots, was very scanty and ill put together. When the young were hatched, I broke through the wall from the inside of the stable to the nest, making the hole large enough to admit a small cage, in which I placed the nest and young; over the inside hole I then hung an old coat, so as to shut out the light from the inside, cutting a small slit in the coat, through which I used to watch the old birds feeding their young within six inches distance. Both birds fed them, at intervals of not more than five minutes. The food consisted almost entirely of centipedes (*Scolopendræ*), with now and then a large spider or blue-bottle fly by way of change. Where they could have found so many centipedes I cannot imagine, as they are insects which lie hid all day under stones &c. The head, in which is supposed to be venom, was always bitten off; and the insect so mangled as to be quite dead. Two of the five young died in the cage, from the old birds not being able to get at them. Of the other three, only one attained maturity, living till October, when, to my great regret, he went the way of all pets. He was very tame, and of most engaging habits and disposition—in fact, what the Spaniards call '*simpatico*.' In his early days he was fed on bread and bruised snails; later on he had more fruit, which I have no doubt killed him. They are difficult birds to keep alive, and (I have since been told) require to be fed on chopped liver.

"The Blue Thrush very often perches on trees, and at Gibraltar and Tangier is frequently seen on the house-tops, though generally observed on bare rocky ground. It is sometimes found in wooded parts if there are any high rocks; for instance, a pair nest at the first waterfall at Algeciraz, which is in the midst of a dense forest. It has a habit in the courting-season of flying straight out from a rock, and then suddenly dropping with the wings half shut, like a Woodpigeon in the nesting-time. The Blue Thrush is very fond of ivy-berries and all fruit."

It seems that they nest more than once a year, as on the 25th of April Mr. Stark found a nest with young about a week old, and on the 3rd of May a nest with five eggs hard sat on, the one in my stable being hatched in the end of June. One set of eggs obtained by Mr. Stark were of the usual delicate pale blue colour, but marked with small russet spots at the large end, somewhat like eggs of the Black Wheatear (*Dromolæa leucura*).

All the nests built on ledges of rocks and open to view are larger and better-built than those placed in holes.

68. MONTICOLA SAXATILIS, Linn. The Rock-Thrush.

"Is found on passage only near Tangier, crossing to Europe in April. Is a scarce species, and very rarely obtained during passage."—*Favier*.

I saw several near Tangier on the 16th of April, 1872, and one on the 30th of March, 1874, also numbers passing at Gibraltar on the 4th of April, 1870; and one was seen there returning on the 26th of September, 1868.

69. CINCLUS ALBICOLLIS, Vieill. The Dipper, or Water-Ouzel.

Occurs in the streams of the Sierras, and is resident. I have shot it near the waterfall beyond Algeciraz, where Mr. Stark found a nest about the 17th of May; but they are not abundant anywhere.

Not yet recorded from Morocco.

70. IXOS BARBATUS, Desfont. The Dusky Bulbul.

Moorish. Bou lág-lág.

"Is very abundant and resident around Tangier. When the oranges are ripe, they are always to be heard and seen chattering and fighting in the gardens. They nest in May,

June, and July, laying from three to four eggs, which are very thin-shelled and tender, of a greyish-white colour, marbled or spotted with reddish spots of two or three shades of brown and purple. The nest is built in the branches of fruit-trees (orange-, apricot-, pear-, &c.), and is shaped like those of the Woodchat Shrike, coarsely interlaced outside with ends of small roots and with creeping plants. They feed on all kinds of fruit and different flowers, are very fond of oranges, and prefer them to any thing else.

"This species is subject to variations, as I have seen two which had the head, breast, and neck brown, with white spots, while the wings, back, and tail were brownish red, the rest being dirty white."—*Favier*.

In accordance with Favier's statement, I found this Dusky Bulbul or *Ixos* in great plenty about the gardens just outside Tangier. They were shy; but one day in March I stalked up to and watched for some time a lot of seven or eight in the Belgian Consul's garden. They were squabbling and playing with one another on a Persian lilac or common bead-tree, the seeds of which they were pecking at; and they reminded me much of some of the Indian Babblers (*Crateropus*), particularly in their flight and garrulous chattering. Besides this noise they have a melodious whistle, which I took down at the time and tried to note thus—*Pwit, Pwit, Qŭitĕrā, Qŭitĕrā,* rather in the tone of a Blackbird. This song (if it may be so called) and their chatter are so remarkable as to attract attention at once.

I took a good deal of pains to ascertain the correct local Arabic name, which is " Bou lág-lág." As no one could tell me the meaning of the latter part, I conclude it is suggestive of their cry, or rather clacking: one of the Arabic names of the White Stork is "Bou lák-lák," from the clacking of their bills. Among the Jews who speak Spanish, they go by the name of "Naranjero" (*litt.* "The orange-man"), from their orange-eating propensities. They make a small hole in the side of an orange and completely clean it out, leaving nothing but a shell of orange-peel, which remains hanging on the tree. I have more than once pulled these husks down, taking them to be sound fruit. Owing to the mischief they thus do, they are not favourites, and consequently are more timid near Tangier than about Larache, where I shot some of them.

I was informed that they do not nest till the end of May, and so had no opportunity of studying their nesting-habits. In the end of April, near Larache, they were evidently not then nesting; and, as at Tangier, all those which I saw were near gardens and villages.

This Bulbul certainly does not occur in the western part of Andalucia; I have tried everywhere for it. If found anywhere, the coast near Tarifa would be the most likely ground; but in the orange-groves there, the Spaniards, when I asked if there was a bird like the "mirlo" which ate oranges, simply looked at me as if I was more "loco" than the generality of "los Ingleses" (who, in their opinion, are all mad), and disclaimed any knowledge of a "naranjero" in the shape of a bird of such size. The Great Titmouse, however, they say eats oranges.

71. CRATEROPUS FULVUS, Desfont. The Algerian Babbling Thrush.

This Babbler is mentioned by Mr. Drake as occurring in the southern part of Morocco, but does not appear ever to have come under Favier's notice in the northern part.

72. SAXICOLA ŒNANTHE, Linn. The Common Wheatear.

Spanish. Culiblanco, Ruiblanca: but these names apply to all the Wheatears.

"This is the most common of the "Traquets," except the Stonechat and Whinchat, but is only seen near Tangier on migration in small flights during March and April, returning in September."—*Favier.*

It is abundant, but seen only on passage, in Andalucia. First noticed on the 4th of March in 1870; a single male bird at Tangier, on the 26th of March in 1874; many seen near Alcala del Rio, on the 4th of April; again passing in numbers at Gibraltar, on the 12th of April. Wheatears were plentiful near Casa Vieja at the end of October and the first part of November, being last seen on the 13th of that month. I have a note also of observing six or seven in the middle of the Bay of Biscay on the 9th of October, when they settled on the steamer, keeping with us till night.

73. SAXICOLA ALBICOLLIS, Vieill. The Eared Wheatear.

According to Favier, this bird is less common than *Saxicola stapazina* near Tangier, but is met with in the same way. Near Gibraltar they appeared to me to be the most frequent, and were first seen there on the 3rd of April, 1870, when I noticed several; and on the 15th of March, 1872, I saw one. I noticed a single bird at Tangier on the 14th.

It nests on the Queen of Spain's Chair, laying about the first week in May. They build a loosely constructed nest among stones and rocks, very often in the same situations as the Blue Thrush. The eggs are light blue, with a zone of brown spots at the large end.

74. SAXICOLA STAPAZINA, Lath. The Russet Wheatear.

"Passes near Tangier during March and April, returning in September. Is the most frequent after the Wheatear, with which bird they travel."—*Favier*.

Is in Andalucia apparently less common than the Eared Wheatear, perhaps because they frequent higher ground; at least I have noticed them more about mountain-tops. They nest about the same time as that species, which they resemble in habits, nest, and eggs.

75. DROMOLÆA LEUCURA, Gm. The Black Wheatear.

Spanish. Sacristan.

This bird is merely named as occurring near Tangier in Favier's MS.; and though it does occur in Morocco, I did not see any. On the Spanish side it is a common and conspicuous bird at Gibraltar, where it is to be seen throughout the year; elsewhere it is migratory, arriving in March, and only found on bare rocky ground. The nest is sometimes in clefts of rocks, so deep in as to be unobtainable. I knew of two or three at Gibraltar in various years.

Mr. Stark took a nest on the 25th of April, near Gibraltar,

containing four beautiful blue eggs hard set on, marked with a zone of light reddish brown spots. The nest was very large, loosely built with grass and heather-roots, lined inside with finer grass, two or three feathers of the *Neophron*, and one bit of palmetto fibre.

76. PRATINCOLA RUBETRA, Linn. The Whinchat.

Moorish. Erdan (*Favier*).

"Is only a passing migrant near Tangier, crossing to Europe in April and May, returning to winter further south in September and October. Is the most common of the Chats, except the Stonechat."—*Favier.*

The Whinchat is met with as above, on the Spanish side being first noticed on the 7th of April; on the 20th, in 1870, many hundreds passed at Gibraltar, as on the 12th of the same month in 1872. I have seen it as late as the 3rd of May. They return in September, and are never seen in winter.

The male has the basal half of the tail white, except the two central tail-feathers, and, from being not so conspicuously coloured as the Stonechat, is less likely to be observed.

77. PRATINCOLA RUBICOLA, Linn. The Stonechat.

Moorish. Bou-erdan. *Spanish.* Caganchina.

"The Stonechat is resident and most abundant about Tangier, being seen in all directions, perched on the tops of plants, bushes, and hedges. They nest from March to July. Some arrive from Europe in September and October, leaving in February and March."—*Favier.*

It is everywhere seen in Andalucia, is one of the most common and at the same time conspicuous birds. They increase in numbers in autumn and spring, but are as common in winter as in summer. I found a nest with five eggs hard set on the 10th of March. There is no doubt that they breed more than once in the season.

Family SYLVIIDÆ.

78. PHILOMELA LUSCINIA, Linn. The Nightingale.

Moorish. Moui el hasin (Prince of Beauty, *Favier*), Umm el hasin (Mother of Beauty). *Spanish.* Ruiseñor.

"This bird is very common around Tangier, arriving during March and April, passing on across the Straits to return in August and September. Great quantities remain to breed about the thick bushy places, chiefly constructing their nest with the fibres of the palmetto, the same material used by the Arabs in making their tents."—*Favier*.

The Nightingale is equally abundant on the Spanish side. The number heard singing in the Cork-wood is perfectly surprising, every clump of bramble-brakes having its pair, though in some seasons they are more numerous than in others; but there are always a great many. They are to be heard at Gibraltar for about ten days or a fortnight after their arrival, but nearly always pass on, though they have been known to nest, as in 1871. My earliest dates of their arrival are the 8th of April 1868, 2nd of April 1869, 7th of April 1870, 1st of April 1871, 21st of March 1872 (Tangier), 30th of March 1874 (Tetuan). The majority arrive about the 12th of April.

They begin to lay about the 1st of May, and usually nest on the ground; but sometimes the nest is placed in ivy or rubbish some two or three feet high. In swampy jungles it is built at the bottom of a bush, and has the lower half constructed of dead leaves, the upper part being made of dry sedges, like that of Savi's Warbler (*Acrocephalus luscinioides*); only it is much more neat, and lined with fine grass, hair, and occasionally I have seen feathers used. As a rule there are young Nightingales in the Cork-wood by the 24th of May.

79. RUTICILLA PHŒNICURA, Linn. The Common Redstart.

Moorish. Houmera (*Favier*).

"This Redstart is only found on passage near Tangier, crossing the Straits in March and April, returning in September and October. It is not so common as *Ruticilla titys*, and is seldom seen settled on rocks."—*Favier*.

The Common Redstart is seen in great numbers near and at Gibraltar on passage. My earliest dates of arrival noticed were the 4th of April 1868, 5th of April 1869, 22nd of March 1870, 28th of March 1874 (Tetuan). In

1872 they passed in great quantities on the 12th, 13th, 14th, and 15th of April, the last noticed being on the 26th, in the Cork-wood. They never appear to remain and nest.

80. RUTICILLA TITYS (Scop.).   The Black Redstart.

Spanish. Colirojo.

" This species is the most common Redstart about Tangier, remaining throughout the winter among rocks and old buildings. They arrive during October, and depart in March. The old birds are solitary; but the immature birds keep together. They shake their tails incessantly, and, holding their heads erect, are difficult to get a shot at " (!).—*Favier.*

The Black Redstart is seen at Gibraltar, as at Tangier, arriving in November, and never being seen after March. They nest, however, a little way north of San Roque.

A specimen I killed at Gibraltar had been eating very small ants.

81. RUTICILLA MOUSSIERI, Olph-Galliard.   Moussier's Redstart.

Under the synonym of "*Ruticilla erythrogastra*" (!!)— a large eastern Redstart—Favier, in his notes, has included Moussier's Redstart; but his description fully identifies it as a male *R. moussieri*. He mentions one killed in 1848. I obtained an adult male at Tangier on the 14th of March, 1872, and saw three others killed in that month in 1874. Although *seeing* a bird is not sufficient evidence to record it unless it be actually obtained, I am sure I saw one in October close to Tarifa, at the time thinking it was a variety of the Stonechat as I passed it by on the road.

82. RUTICILLA WOLFII (Brehm).   The Blue-throated Warbler.

Spanish. Soldiya.

" Found near Tangier only on passage, and then very rarely. I only obtained them four times—in 1839, 1844, 1866, and 1867. They cross to Europe in February and March, returning in October."—*Favier.*

I imagine this species must pass further to the east, as about Gibraltar I only saw one alive, which I shot on the 1st of March, as it was perched on some rushes in an old "salina" near Palmones; another, shot in November 1873, at the same place, is in the possession of Lieutenant Reid. I have seen specimens from Seville and Granada.

The adult males have a white spot in the middle of the blue throat.

83. ERYTHACUS RUBECULA, Linn. The Robin.

Moorish. Humar sidri. *Spanish*. Petirojo.

"Is resident near Tangier, and very common in all the gardens around the town. Numbers also migrate, arriving during October and November, departing in February and March."—*Favier*.

Common throughout Andalucia in winter. The Robin only comes to Gibraltar from about the middle of October to the middle of March, but then in considerable numbers. They are resident in the Cork-wood, nesting abundantly in April, where, one day in May, my attention was attracted by the chattering and scolding of two Robins, evidently in a great state of alarm and excitement; close by them was a palmetto bush, to and from which they were flying, hovering over it, but not settling. At first I thought a cat, or perhaps an ichneumon, was lying up; but on peeping quietly into it, I saw a snake, some three feet long, in the act of swallowing a half-fledged Robin at the edge of its nest. I drew back a pace, and fired a small charge of dust-shot into the reptile's head, cutting it nearly in half. The brute, however, had disposed of all the young birds; so, though too late to save them, the parents were rescued, as no doubt they would have shared the fate of their progeny but for my interference.

I hung the snake up in the nearest bush, "*pour encourager les autres*," the old Robins all the time watching my proceedings; and I hope they were able to understand that their loss was partially avenged. The quantity of young birds— Robins, Nightingales, and similar ground-nesting birds— which are destroyed by snakes must be very great.

84. ACCENTOR COLLARIS (Scop.). The Alpine Accentor.

I have only seen this bird at the back of the Rock at Gibraltar in winter. I shot one on the 1st of February, and saw others on the 26th of the same month in 1870. Mr. J. H. Gurney, jun., who was passing through Gibraltar, was the first to notice it at the signal-station. There was a specimen in an Instituto at Seville; and no doubt it is found on all the high rocky ground, though I could not meet with it on the Sierra del Niño or elsewhere.

85. ACCENTOR MODULARIS (Linn.). The Hedge-Accentor.

M. Favier did not include this bird in his list of Moorish birds, merely mentioning it as occurring near Gibraltar, having met with it during his " triste séjour" in that place, in November. I have seen specimens from the African side of the Straits. On the Spanish side it is found in winter. I have shot it in the Cork-wood in January; but it is not common.

86. SYLVIA SALICARIA, Gm. The Garden-Warbler.

"Found near Tangier, on passage to Europe, in April and May, returning in October, when it is nearly as plentiful as the common Whitethroat."—*Favier*.

The Garden-Warbler mostly arrives during the middle of April. I first observed one on the 10th. The latest I saw was on the 7th of October. They nest around Tangier and in the Cork-wood, laying about the 10th of May, and are brought into the market at Gibraltar as "becafigos;" later in the season, like most of the genus, they are great devourers of figs.

87. SYLVIA ATRICAPILLA, Linn. The Blackcap.

"Is nearly as common as *S. melanocephala* about Tangier, being seen on all sides during migration, passing north in January and February, returning in October. Many remain to nest."—*Favier*.

The Blackcap is to be seen during every month in the year, but is, of course, most common in February and October. They sometimes nest on the Rock, always plentifully in the Cork-wood. I have seen the young fully able to fly on the

24th of May. They chiefly fed in my garden for some months on the seed of the so-called " pepper-tree " (*Schinus molle*), in company with the Black-headed Warbler, and, to my surprise, with the Black Redstart; at least I saw the latter pecking at the seeds.

The young males in winter have brown heads, like the females. The species may be distinguished at a glance from the other black- or dark-headed Warblers likely to be met with near Gibraltar by the absence of white on the tail.

88. SYLVIA ORPHEA, Temm. The Orphean Warbler.

Andalucian. Canaria.

" This Warbler passes by Tangier in April and May to return in September, travelling in company with the White-throats. Is not common, and in some years scarcely met with."—*Favier*.

I could not find the Orphean Warbler to be common near Gibraltar. I never had a specimen brought to me; nor did I succeed in getting it till the 17th of May, 1871, when I found a nest on a branch of a pine tree in the " Second Pine-wood." I shot both the old birds, which were very fearless, particularly the female, who contained an egg ready for exclusion. The nest had only three eggs in it, and was badly built, being composed of grass and lichens. I imagine that this Warbler must chiefly pass further to the east. They nest around Seville and are common about Madrid; but I could not ascertain that they bred near Tangier.

It is at once distinguished from *Curruca atricapilla* by the tail, which has the two outside feathers on each side tipped with white, and the exterior web also white. The legs are bluish, the irides yellow. The bird is also slightly shorter than the Blackcap.

89. SYLVIA MELANOCEPHALA, Gm. The Black-headed Warbler.

Moorish. Shorrir (*Favier*).

" This Warbler is resident and very abundant near Tangier; some migrate, crossing the Straits during February and

March, returning in September. They are to be seen everywhere, nesting in small thorny bushes. The nest is not well-built, and is made of strips of plants and blades of grass, without roots; rarely there is a little wool. It is lined with the down of some cotton-like plant, fine fibres of roots, and a few horse-hairs. They lay from April to July."—*Favier.*

The Black-headed Warbler, equally common around Gibraltar, is found in all scrub, gardens, and in the midst of woods, scolding with a chattering noise much like that of our common Wren. It might well be named the Gibraltar Warbler, being the only species which is a regular resident on the Rock. In habits it much resembles the Blackcap, but is more restless and obtrusive, and consequently more conspicuous; the contrast between the jet-black head of the adult male and the white throat also renders it more liable to be noticed. There were in different years several nests in my garden, which I religiously preserved; but what with cats and inquisitive human beings, they seldom succeeded in rearing their young. The earliest egg laid was on the 12th of March; this was built in a small rose-bush, and was spoiled by a gale of wind, which blew all the eggs out of it, being the only one I ever saw in what could be called an open bush. All the others were placed in thick bushes, generally box, about two to four feet from the ground, and were formed of grass with a few bits of cotton-thread, lined with hair. The eggs vary in number from three to five. The male assists in incubation.

This Warbler is, like the Blackcap and Garden-Warbler, very fond of figs and grapes and all kinds of fruit. The feathers at the base of the bill and the throat are often much coloured with the pollen of cactus and aloe flowers, and with the seed of the " pepper-tree."

The adult male has brick-red eyelids.

90. SYLVIA CURRUCA, Lath. The Lesser Whitethroat.

This species is, as far as my observations go, rare. I obtained it once in my garden at Gibraltar in April, and another on the 19th of April, 1872. It does not appear to have been observed by M. Favier.

91. SYLVIA RUFA (Bodd.). The Common Whitethroat.

"Arrives about Tangier and crosses to Europe in April and May, returning to winter further south in September and October. Is nearly as abundant as the Blackcap, and seen on passage in small flights. On their return they have the top of the head the same colour as the back, like the females in spring."—*Favier*.

I never saw the common Whitethroat near Gibraltar in winter. I noticed their first arrival in 1870 on the 7th of April, in 1871 on the 7th of April, in 1872 on the 11th of April, many passing on the 19th and 20th, and in 1874 on the 8th of April. They nest abundantly in the Cork-wood, and elsewhere in quantities in marshy places, building their nests in thick leafy plants, often in those of the willow herb (*Epilobium*); the average time for their laying is the 7th of May. When looking for Savi's Warbler, I sometimes found a dozen nests in the day.

92. SYLVIA CONSPICILLATA, Marm. The Spectacled Warbler.

Favier merely says of the Spectacled Warbler, that about Tangier it is not common, and only seen on passage north in March. He gives no date of its autumnal migration, but states that they pass the winter somewhere further south.

The earliest date on which I obtained it near Gibraltar was on the 10th of March. It is a conspicuous, scrub-haunting bird, frequenting dry and more open ground than the Whitethroat, being seen among cactus bushes. A sure place for finding them is in the Cartcian hills. They remain during the nesting-season, but I did not myself find the nest.

The irides are very light brown; the inside of the mouth pale yellow.

93. SYLVIA SUBALPINA, Bonap. The Subalpine Warbler.

Favier merely states that "this species occurs near Tangier on passage in March and April, and again in October." It is not often noticed near Gibraltar; but I shot one on the 20th March, 1870, and at Tangier on the 26th of March and 27th of April, 1874. On the 27th of March, 1871, I saw eight or ten among the flowers and trees on the Alameda de Apodaca at Cadiz; they were exceedingly tame, and I watched them for a long time hopping about in and out among the flowers like a common Wren. One or two were very bright-coloured males. I also saw this Warbler on the 25th of April, 1869, in the Coto del Rey. Lord Lilford informs me he found a nest early in May, built in a gum-cistus bush in the Coto del Donaña, the eggs being very hard sat-on.

I shot one on the 3rd of May, which evidently had a nest, though after spending some hours looking for it, I was unsuccessful in finding it.

I never had the good fortune to discover a nest; but they are said to build much in the same manner and situations as *S. melanocephala*, the eggs being also very similar to those of that bird.

This species is very apt to be confounded when flying and hopping about with *Melizophilus undatus*; but the length of the tail, shorter in proportion, distinguishes it from that bird.

94. MELIZOPHILUS UNDATUS (Bodd.). The Dartford Warbler.

Spanish. Colorin, Caganchina.

"Is resident but not abundant near Tangier. Some migrate to Europe in March, to return in August. It is solitary in habits. They make a clumsy nest of grass and roots, lined with very fine coils of palmetto-fibre, laying in April."—*Favier.*

The Dartford Warbler is resident and not uncommon in all the scrub-covered hills on the coast near Gibraltar, particularly about San Roque, but is most abundant during the breeding-season on the sides of the sierras, nesting in the heather about the 8th of April, on which date Mr. Stark found a nest near Algeciraz with three eggs. There is no doubt they nest at Gibraltar, as they occasionally remain there through the summer.

95. PHYLLOSCOPUS SIBILATRIX, Bechst. The Wood-Warbler.

This species is not mentioned by Favier as occurring near Tangier, where, however, it is found, though not commonly. On the Spanish side of the Straits it is the scarcest of the four species of *Phylloscopus*. It was first seen on the 22nd of April. I have killed it in my garden at Gibraltar, and some remain during the nesting-season in the Cork-wood; but I have not observed it in winter, and I was unable to notice the date of its departure south.

The nest is not lined with feathers, like those of *P. rufus* and *P. trochilus*, and is always placed on the ground.

The bird is easily distinguished by the streak of bright yellow over the eye, and the white colour of the underparts, and is also the largest species of the genus which occurs near Gibraltar.

96. PHYLLOSCOPUS TROCHILUS (Linn.). The Willow-Warbler.

Moorish. Simriz.

"The most common of Willow-Wrens near Tangier; crosses the Straits in April, returning in November."—*Favier.*

There is no doubt, although I did not find a nest, that this species nests near Tangier. In the vicinity of Gibraltar they are to be found throughout the year in the Cork-wood, where they breed. I have seen the young able to fly on the 8th of May. Although universally distributed in winter, they are most common on passage in March and October.

97. PHYLLOSCOPUS BONELLII, Vieillot. Bonelli's Willow-Warbler.

" Found during migration near Tangier, in company with *P. rufus* and *P. trochilus*, but is not so numerous. They return in September."—*Favier.*

This species, about the size of the Chiff-chaff, is found in plenty near Gibraltar, nesting in the fern in the Cork-wood; the earliest I noticed arriving was on the 1st of April. It is easily recognized as a species by the white of the underparts, by the yellow shade on the rump, and by the white streak from the angle of the beak to the eye over the ear-coverts. Furthermore it never occurs in the winter months.

98. PHYLLOSCOPUS COLLYBITA (V.). The Chiff-chaff.

According to Favier this bird is nearly as common as *P. trochilus*, crossing to Europe in February, March, and April, returning in October and November.

The Chiff-chaff is to be seen throughout the year in the Cork-wood, but is most common from November to March. I found a nest on the 21st April in a bush about six inches from the ground.

These four species of *Phylloscopus* all build domed nests, usually on the ground, but occasionally in bushes or fern at an elevation of sometimes two feet or more above the ground; this is particularly the case with Bonelli's Willow-Warbler.

The Chiff-chaff is difficult to tell from the common Willow-Warbler, but is always smaller and the legs are darker, being almost black, the eyebrow is not so well defined, and it is a more dull-coloured bird than *P. trochilus*. The note is also very different and distinct. The young of all the species are more highly coloured than the adult birds; but the genus is a very troublesome and perplexing one to the student, and only to be elucidated by observing the different species in a wild state. The skins shrink and the colours fade so much that a cabinet naturalist is much the most puzzled with them.

99. HYPOLAIS POLYGLOTTA (Vieill.). The Yellow Willow-Warbler.

"Arrives and crosses to Europe in April, returning in August and September, many remaining to nest around Tangier."—*Favier*.

This Willow-Warbler is exceedingly plentiful near Gibraltar, being one of the latest of the spring arrivals; the first I observed was on the 25th of April, and the earliest date on which I saw eggs was on the 14th of May. The birds frequent trees and bushes, especially willows and sallows; and the nest, neatly built and cup-shaped, is in a great measure composed of sallow-cotton and thistle-down; it is placed in bushes, and usually contains four pinkish-tinged eggs, marked with blackish spots.

This bird was figured by Yarrell as the Melodious Willow-Warbler (*H. icterina*), a slightly larger species. The present bird is unknown in England, and has the first primary small, but longer than the primary coverts; in *H. icterina* it is scarcely as long.

The inside of the mouth of this species is bright orange-yellow.

100. HYPOLAIS OPACA, Licht. The Western Pallid Warbler.

"This Warbler is nearly the same in size as *H. polyglotta*, but is somewhat larger, and is identical with that bird in habits, times of arrival and departure, and also in manner of nesting. They build on trees, bushes, and small plants, laying in May or the beginning of June."—*Favier*.

The above was under the head of *H. elaica*; but as that bird does not occur and all Favier's specimens belong to *H. opaca*, there can be no doubt that the notes refer to the latter species.

This Warbler is the latest of all the spring migrants that arrive in Andalucia, being a little later than *H. polyglotta*. It is much more plentiful eastward of Gibraltar than in the immediate vicinity, where it is rare. Another species, *H. olivetorum*, is stated to have been met with at Tangier and Fez, but did not come under my observation.

101. CISTICOLA SCHŒNICOLA, Bp. The Fantail-Warbler.

Moorish. Bou-fesito (Father of eloquence). *Spanish.* Cierra-puño, Tin-Tin.

"Is the most common of the aquatic Warblers around Tangier, and seen migrating in lots of from ten to twelve during March and April, returning in October, November, and December. Many remain to breed, nesting twice in the season."—*Favier*.

This diminutive Sedge-Warbler, as I may call it, is resident near Gibraltar, and exceedingly plentiful, in the winter frequenting marshy ground wherever there is any herbage, such as grass, sedges, or short rushes. In the spring they go to the corn-fields as well, never, however, being found away from water. I do not recollect ever seeing them perch on a bush or tree, but always on some plant. Their note and jerky flight somewhat remind one of the Meadow-Pipit; during the nesting-season in particular they will fly darting about high over head for several minutes, continually uttering their squeaky single note (whence the name of Tin-Tin), all the time evidently trying to decoy the intruder from their nest. They undoubtedly breed twice a year—according to the Spaniards, three times. I have found the young well able to fly, and a nest with eggs ready to hatch, on the same day, the 19th of April, an unfinished nest on the 8th of May, and a nest with eggs very hard sat-on on the 10th of that month.

The nest much resembles the cocoons which are so common on pine trees in some parts of Spain; any one would take them for the web of some insect; but they are very troublesome to find. They are made of the cotton of plants and thistle-down, with small bits of grass beautifully sewn and interwoven with the corn or grass in which the nest is built; the entrance is at the top, the bottom being the broadest part, the whole length about five inches. The usual number of eggs is five, generally of a pale blue; but, as is well known, they vary strangely in colour.

The inside of the mouth black; the irides very pale brown.

102. AEDON GALACTODES (Temm.). The Rufous Warbler.

Moorish. Houmira. *Spanish.* Alzacola, Rubita, Viñadera.

"Abundant in the vicinity of Tangier, arriving in April and May, returning during September, many remaining to breed. Their habits are the same as those of the Nightingale. The nest, large and well built, is placed at some height from the ground, in thick foliage. The eggs, from five to six in number, only differ from Sparrow's eggs in the spots being more reddish. The males assist in incubation."—*Favier.*

On the Spanish side, this "Cocktail" Warbler, as I should call it (from its well-known habit of continually jerking its tail up), is very plentiful, frequenting sandy lanes hedged with aloes and prickly pears, such as those close to the First Venta, near Gibraltar. As Favier remarks, they resemble the Nightingale very much in their habits, and are at first sight very likely to be mistaken for it; only the Nightingale comes some three weeks or a month earlier.

The Rufous Warblers mostly arrive near Gibraltar between the 1st and 5th of May. The earliest I noticed in 1869 was on the 28th of April, in 1870 on the 29th, in 1871 on the

22nd, and in 1872 on the 28th of that month, the migration lasting quite for weeks. They nest about the last week in May.

In places where there are many vineyards (which they frequent) they are known as *Viñadera*. *Alzacola* is the local name about Gibraltar; and "Cocktail" is very nearly a translation of it. "Rufous *Sedge*-Warbler," as this bird has been hitherto called, is a most inappropriate name, as they are never seen near either water or sedges.

103. CETTIA SERICEA, Natt. Cetti's Warbler.

Favier states that "this Warbler is rare near Tangier, and seen on passage in February and March, to return in October." This, however, is quite different from my own observations. They certainly are not rare in spring near Tangier, where, as on the Spanish side, wherever there are thick bushes (generally bramble-brakes close to water) Cetti's Warbler is to be heard. Perhaps many migrate; but at Casa Vieja they are quite as common during the winter months as at any other season, and, somewhat like our own Robin, may be heard singing at all times. Very difficult to see in the breeding-season, in the winter months they do not skulk so much. They are excessively restless, being ever on the move; and often in the winter, when hidden up in the sotos near Casa Vieja, have I watched them quite close to me; but the slightest movement on my part sent them off to the thickest depths of the jungle. In the breeding-season it is almost impossible to catch a glimpse of one.

The only chance of shooting them is at the nest, which is always placed some distance from the ground, generally at a height of about two or three feet, and is either situated in a thick bush or (when in a bushy swamp) constructed, somewhat like the nest of the Reed-Warbler, on the stalks of reeds and *Epilobium*. These nests are built of bits of small sedges, intermingled with willow-cotton, and chiefly lined outside with strips of the stems of the *Epilobium*, inside with fine grass, a few hairs, and bits of cotton at the top. Those nests built in bushes are chiefly constructed with grass and cotton, and are entirely lined with hair. All the nests are deep and cup-shaped, largest at the base, measuring about 4½ inches in height, the inside depth being 2¼, the internal diameter 2¾ inches. The beautiful pink eggs, laid about the end of April, are usually five in number; but I have known only three. They lose much of their beauty when blown.

The birds are rather irregular as to the time of nesting, as I have seen nests nearly on the point of hatching and others with fresh eggs on the same day (13th of May).

The males are about a quarter of an inch longer than the females; but the plumage of the sexes is the same. Sometimes the tail is marked, like Savi's Warbler, with indistinct bands of dark brown. The inside of the mouth is yellowish. The number of tail-feathers is *ten* only.

104. Acrocephalus schœnobænus (Linn.). The Sedge-Warbler.

Favier says this is a very rare species near Tangier, that he seldom saw more than one or two on passage, in March or in September. I never obtained it on the Spanish side.

This species has a broad yellow-white eyebrow, the top of the head being brown streaked with blackish brown.

Length about 4¾–5 inches.

105. Acrocephalus aquaticus (Lath.). The Aquatic Warbler.

This species is mentioned by Mr. Drake as having been met with in Morocco. It is found in Andalucia, though I only saw it once myself.

The top of the head, which is brown, has a buffish white band down the centre and a buff-coloured eyebrow, and is a rather smaller bird than the Sedge-Warbler.

106. Acrocephalus nævius (Penn.). The Grasshopper-Warbler.

Is recorded from Morocco and also from Andalucia in winter; but I did not myself observe it.

Length about 5½ inches.

107. Acrocephalus luscinioides (Savi). Savi's Warbler.

Recorded by Mr. Drake as met with in Morocco.

I only found it in one locality in Andalucia, where once (in winter), when snipe-snooting, I noticed some old nests in the sedges, which apparently belonged to this species, and made up my mind to try the next spring for them. However, for two years I was unable to do so; but in 1874 I went to this place in May with two friends, Mr. Stark being one, and we succeeded in finding thirteen nests, nine of which fell to my share. The first nest was found by Mr. Denison, on the 4th of May, and contained four fresh eggs; the others as follows:—on the 6th, one nest with four fresh eggs; on the 7th, three nests—one empty (deserted), two with four eggs each, one lot fresh, the other hard sat-on; on the 8th, one nest procured with three eggs slightly sat-on, and one nest with five fresh eggs; on the 9th, two nests with four eggs each, all hard sat-on, and one nest with three young fully fledged; on the 11th, one nest with five fresh eggs; and on the 13th, one nest with two fresh eggs.

By this it will be seen that the time of their breeding is rather variable. I do not like to give the name of the exact locality where these birds nest, as (owing to the detestable system of amateur dealing and making money out of the eggs and skins of scarce birds) I am afraid Savi's Warbler would suffer heavily.

The precise time of their arrival I could not ascertain; but it is after the 6th of March; and they are all gone by September. The nests, sometimes very near to one another, are most difficult to find, and were, without exception, built in places where the mud and water varied in depth from two or three inches to perhaps two feet. All but one were in sedges, so well concealed as only to be found by accident. I spent sometimes the whole day in these marshes, looking in vain, with my gun in one hand and a sickle in the other, which I used to open the sedges with, as it cut one's fingers severely to try and move them with the hand. What with the hot sun and the stink of the mud, I used to despair utterly after hours of fruitless search, but generally found a nest in the evening. The whole marsh was trodden down by us as if a herd of cattle had been in it; but perhaps the next day, going over the same ground, one would find a nest in a bunch of sedges which had been passed by within a yard. The nests were all alike, loosely and clumsily built, solely constructed of dead sedge, often placed so close to the water that the base was wet; they were always in the open marsh, none, that I saw, under bushes or in tall rushes or reeds.

The single nest that was not in sedges was in a tuft of the spiky rush so common in wet ground. In this case (the first one, found by Mr. Denison) the bird flew off—the only instance in which it did so, as they creep off generally like a mouse. On one occasion I cut away all the sedge round the nest, except just the patch in which it was built, as I wanted to shoot the bird from the nest to make certain of the identity of the eggs; but even then, after watching the old bird go in to the nest she would not fly off, but ran across the open space which I had cut away till she gained the shelter of the uncut sedges. Much more frequently seen than Cetti's Warbler, the great difficulty is in finding them when shot. If killed on the wing, it is almost hopeless to look for them; and those that I did obtain I have to thank my dog for finding, though he did spoil one or two. They are most easily to be got in the morning and evening, when the male perches on a sallow bush or tall reed and sings his grasshopper-like song, or rather whir.

I only found them in one particular locality: in other marshes, very similar in appearance, I failed to hear or see them; and they probably require a very large extent of sedge.

The eggs are of a whitish ground-colour, marked all over with minute spots of brown, thicker at the larger end, often forming a well-marked zone. Sometimes the ground-colour is buff; but I only saw two or three of this hue.

The head, wings, and tail are reddish brown, the tail indistinctly barred with darker bands of brown, and cuneiform in shape. The length is about 6 inches, extent from wing to wing $7\frac{1}{2}$. Legs and feet pale brown, the claws darker; irides olive-brown; upper mandible dirty white, with dusky tip; lower one blackish; inside of mouth of adult pale salmon-colour, that of young bright yellow.

108. ACROCEPHALUS STREPERUS (Vieillot). The Reed-Warbler.

Not mentioned by Favier as occurring in Morocco; nor does Mr. Drake appear to have observed it in that country.

On the Spanish side it appears in spring. The exact date of arrival I could not ascertain; but it is somewhere about the end of March. I never met with the Reed-Warbler during the winter months; but in the marshes at Casa Vieja, about the first week in May, I found it breeding in abundance. It keeps among the sallow-bushes, but builds its beautiful nest suspended on the dead stems of the *Epilobium* or willow-herb, which grows in luxuriant tufts in the swampy jungle. These nests are constructed externally of strips of the rind or peel of the dead *Epilobium*-stems interwoven with sallow-cotton, the interior being composed of fine grass lined with the same material. The usual number of eggs was four, of a pale greenish colour, marked all over with ashy spots.

The upper part of the plumage is a uniform pale brown; chin and throat white; the rest of the underparts pale buff, darkest on the flanks; legs dark brown.

Length from 5 to 5·5 inches; wing 2·5, tarsus 0·9.

109. ACROCEPHALUS ARUNDINACEUS, L. The Great Sedge-Warbler.

This large species, though it occurs in Morocco, is not included in Favier's list. Exceedingly plentiful in Andalucia, arriving in April and chiefly frequenting tall reed-beds. They are very noisy, and, like other aquatic Warblers, conceal themselves at the slightest alarm. However, by ensconcing one's self and remaining quiet among the rushes, they are easily obtained, as they soon come out and sit singing and chattering on the top of some tall reed. They nest late in May, and build a nest resembling that of the Reed-Warbler (*Acrocephalus streperus*), only of course much larger. I have seen them building during the first fortnight in May, picking and carrying away the down of the bulrush to use in constructing their nests. The inside of the mouth is orange-yellow.

Length about 8 inches; tarsus 1·12. The females are slightly the smaller.

110. REGULUS CRISTATUS, Koch. The Gold-crested Wren.

I have never met with this species in Andalucia, where it has been recorded as common; and perhaps the young of the next species may have been mistaken for it. Possibly it may occur irregularly, like the Siskin—that is to say, not in all winters consecutively.

The length is about 3½ inches; wing, carpus to tip, 2.

111. REGULUS IGNICAPILLUS (L.). The Fire-crested Wren.

This species is resident and common in the Cork-wood and in the wooded valleys at the back of Algeciraz, coming as near to Gibraltar as the Malaga Gardens, close to San Roque. They nest rather early, the young being able to fly on the 15th of May.

The distinguishing mark of the adults is the greyish-white eyebrow, which lies between two black lines, one of which is at the base of the crest, the other running from the beak through the eye; below this line is another whitish mark, with a third and more faintly marked dark line,—making three dark lines and two light-coloured streaks, or, as the French have it, *triple bandeau*. The crest of the male is of a fine flame-coloured yellow; that of the female, lemon-yellow.

The size of the two species is about equal.

Family CERTHIIDÆ.

112. TICHODROMA MURARIA, Linn. The Wall-Creeper.

I have never met with this bird; it is recorded from the Sierra Nevada and the north of Spain, and has been stated to have been seen at Gibraltar.

It is a grey bird, and the only species known of its genus; it has the wing-coverts and the basal half of the outer webs of the quill-feathers crimson, the average length being about 6 inches.

113. CERTHIA FAMILIARIS, Linn. The Tree-Creeper.

Spanish. Barba-jelena, Trepa-troncas.

The Tree-Creeper is resident and common in the Cork-wood and in the valleys near Algeciraz, nesting in April. I saw a single specimen shot near Tangier about the 20th of April, the only one I heard of on the African side.

Family TROGLODYTIDÆ.

114. TROGLODYTES PARVULUS, Koch.   The Wren.

Spanish. Cucito, Ratilla.

"Resident near Tangier, and numerous, nesting from March to June. Some are migratory, arriving in November, leaving again in February."—*Favier*.

The above remarks equally apply to the Wren on the Spanish side, where it is most abundant in winter. It nests very early; and I have seen young fully able to fly on the 26th of April. They are resident on "the Rock."

It is very curious that this little bird should be a resident in the scorching sun of Morocco and Andalucia as well as in the bitter cold of the hills in Inverness-shire, where they are one of the very few birds which remain to brave the winter.

Family PARIDÆ.

115. ACREDULA IRBII, Sharpe and Dresser. The Spanish Long-tailed Titmouse.

Spanish. Mito.

This little bird is only to be found around Gibraltar in the cork-wood of Almoraima, chiefly keeping to the sotos and to the district round the Mill, the Long Stables, and the second venta. Similar in its habits to the British species (*Acredula vagans*), the nest and eggs are also exactly the same as those of that bird. I found the young able to fly by the middle of April, and on the 12th of that month found a nest with seven young fully fledged; this would make the date of laying about the 20th of February. The nests, without exception, were all built in the thorny creeper (called *Zarzaparilla* by the Spaniards), which forms regular net- or lattice-work walls from the ground to the lower branches of the trees, and is usually placed about 15–16 feet from the ground. The nests are very difficult to get at, the only way being either to cut or shoot away the creepers above it—often no easy matter. The only eggs which I obtained were addled ones, left in nests from which the young had flown. The adults differ from the British and North-European species in having the entire back bluish grey. The eyelids of the adults are golden yellow; those of the young brick-red.

116. PARUS CÆRULEUS, Linn. The Blue Titmouse.

Spanish. Herrerita.

The Blue Tit is very common in Andalucia, being resident and particularly abundant in the cork-wood of Almoraima, generally nesting (about the middle of April) in the decayed hollow branches of the cork-trees.

Spanish specimens are very bright in colour—one or two so much so that, until I had seen a specimen of *Parus teneriffæ*, I imagined them to be that species.

117. PARUS TENERIFFÆ, Less. The Ultramarine Titmouse.

Moorish. Bou reziza (*Favier*), as also in Algeria (*Loche*).

"Is resident near Tangier, but less frequent in December and January than during other months. They nest in holes of trees, in April, laying from four to five eggs, white, with very small red spots (similar to those of *Parus cæruleus*). This species replaces *P. cæruleus* in Morocco, and appears to be a variety of that bird constant to this climate."—*Favier*.

I have never been able to detect the Ultramarine Tit on the Spanish side; nor have I seen *P. cæruleus* in Morocco, where the present species is plentiful, their habits &c. being identical.

The African bird is easily recognized by its grey back and the deep blackish blue on the crown of the head, as well as on those parts which are cobalt-blue in the European species.

118. PARUS MAJOR, Linn. The Great Titmouse.

Spanish. Quive-vive, Carpintero, Carbonero.

Favier considers this species to be extremely scarce near Tangier. I have seen specimens obtained there in winter, but never personally observed one. On the Spanish side of the Straits it is extremely plentiful, and to be heard wherever there are any trees. Numbers nest, in April, in the holes of the cork-trees. There is an increase of their numbers in winter, when they visit the Alameda and gardens at Gibraltar, being the only Calpeian representative of the genus *Parus*.

119. PARUS ATER, Linn. The European Cole Titmouse.

I never met with this species, which is said to occur in Andalucia, though I saw a specimen in a museum at Seville supposed to be Spanish-killed.

The Algerian Cole Titmouse (*Parus ledouci*) in all probability is to be met with in Morocco. This bird has those parts of the head and nape of the neck lemon-yellow which are white in *P. ater*, the under-surface being also lemon-yellow.

The Marsh-Tit (*P. palustris*), stated by Mr. Saunders as occurring near Granada, I have never met with near Gibraltar or anywhere in Andalucia.

120. PARUS CRISTATUS, Linn. The Crested Titmouse.

Spanish. Capuchino.

The Crested Titmouse is resident and common in the cork-wood of Almoraima, in all the neighbouring pine-woods, and in the valleys and on the hill-sides at the back of Algeciraz up to near Tarifa, wherever the cork-tree grows. It nests (about the 10th of May) in the hollow stumps of boughs of the cork- and pine-trees, the eggs being about five in number, much spotted, and resembling strongly those of the Creeper (*Certhia familiaris*).

I have reason to think it occurs in Morocco, but only mention this with the view of directing the attention of future collectors there, in order that they may look out for it.

Family MUSCICAPIDÆ.

121. MUSCICAPA ATRICAPILLA, Linn. The Pied Flycatcher.

"Very abundant near Tangier on passage, crossing to Europe in pairs and small flights during April and May, returning in September and October."—*Favier*.

In Andalucia the Pied Flycatcher only appears during

migration; and I have never been able to detect it remaining to nest. The earliest date of arrival noticed was the 8th of April; from then till the 1st of May they pass in great numbers, returning late in September. The latest date on which they were observed was the 17th of October 1870, when I killed one in an old owl's cage, where there was a lot of carrion which attracted flies; and, again, in 1871 I noticed them on the 16th of October.

The Collared Flycatcher (*Muscicapa albicollis*) I have not noticed, nor have I seen a specimen from either side of the Straits.

122. BUTALIS GRISOLA (Linn.). The Spotted Flycatcher.
Spanish. Papamoscas.

"This Flycatcher is very common near Tangier, where they arrive in April and May in pairs and small flights, some remaining to nest, the rest passing across the Straits to return in September, when they disappear. Near Rabat they are called *Sorsh* by the Arabs."—*Favier.*

The Spotted Flycatcher is exceedingly numerous near Gibraltar, chiefly nesting in the pine-woods. It was first seen on the 11th of May 1870, on the 3rd of the same month in 1871, and on the 8th in 1874. The first egg obtained was on the 24th of the same month. I regret not to have any note of their departure; but it is previous to the middle of September.

Family HIRUNDINIDÆ.

123. CHELIDON URBICA (Linn.). The House-Martin.
Spanish. Vencejo.

"As common as the Swallow near Tangier, this species is seen in flights on passage, crossing to Europe in February, returning in September and October, frequently travelling in company with *Hirundo rustica,* and, as in their case, remaining to breed in some numbers. They often make their nests touching one another, as many as sixty being joined together; the entrance-hole is sometimes at the side, sometimes in the centre, according to the position of the nest. They are named *Khotaifa* by the Arabs, indiscriminately with the Swallow."—*Favier*.

The above notes equally apply to the House-Martin in Andalucia. The earliest date of arrival noticed at Gibraltar was the 5th of February. They (as well as *Hirundo rustica*) frequently nest on rocks, like *Cotyle rupestris*.

124. HIRUNDO RUSTICA, L. The Common Swallow.

Moorish. Khotaifa. *Spanish*. Golondrina.

" Great flights of Swallows pass in January and February to Europe, returning in September and October to join those which remain near Tangier to nest, all leaving to go further south for the winter. The Moors believe that it offends God to kill these birds, in the same way as they believe it pleases or soothes the Evil One to kill the Raven (*Corvus corax*). The stories on which this superstition is founded are too long to relate; but I was informed by one person that the Swallows and White Storks were inspired by Allah to protect the harvest and the country from noxious insects and reptiles, and that the birds themselves (knowing the benefits they confer on man) ask in return protection for their offspring by building their nests on the walls of towns and houses, and that therefore any one who kills them must be a Kaffir, *i. e.* not a true believer of the Prophet, especially as the birds would only be killed for mischief, being useless when dead."—*Favier*.

I wish this belief could be instilled into the minds of English people, who kill and destroy every rare bird they see, perhaps more through ignorance than love of destruction.

About Gibraltar the Swallow generally arrives about the 13th of February, although I have occasionally seen a straggler in December and January. I have seen them crossing the Straits in considerable numbers up to the 15th of April; the latest I noticed were passing on the 24th of that month. I have observed the nest finished on the 23rd of February, and young birds able to fly on the 24th of May. One of each pair, when they first arrive, is tinged with a rufous buff-colour on the underparts; and as these are slightly larger in size, I think they are the male birds; but I would not, even for the sake of proving this, kill one. I remember, on a very cold day (the 13th of March, 1874), Mr. Stark particularly drawing my attention to this difference in the pairs of birds, which, driven by the cold into the stables and outhouses of the venta at Pulverilla, were sitting side by side, touching one another, allowing us almost to catch them. The contrast in their colour was then most conspicuous; but they appear gradually to lose this rufous tinge as the season advances, and by the end of April it is not apparent.

125. Cotyle riparia (Linn.). The Sand-Martin.

"Migratory, and the least abundant of the Swallows about Tangier, arriving to cross the Straits in March and April, returning in October to disappear for the winter."— *Favier.*

I found the Sand-Martin at Ras-Doura in small numbers, and have no doubt that they were nesting in the vicinity; they nest in the neighbourhood of Seville, but near Gibraltar are only met with on passage. The first seen by me was on the 24th of March 1870, 22nd of March 1871, 24th of February 1872, 28th of February 1874; they were seen passing as late as the 24th of April. On the 13th of May I saw, in the evening, over some marshes near Vejer, a flight of Sand-Martins numbering many hundreds; I might say, thousands. I noticed them on the 14th of October on their southward journey.

126. Cotyle rupestris (Scop.). The Rock-Martin.

"Nearly as common as the House-Martin about Tangier. Sometimes they pass in large flights, crossing the Straits in February and March, returning in October and November." —*Favier.*

The Rock- or Crag-Martin, though universally distributed during the breeding-season in the rocky sierras, is to a great extent migratory. Those which do not quit the country appear during the day-time in low ground near the coast about the middle of October, great quantities being then seen about Gibraltar. They roost at that season about low rocks, especially at Gibraltar. In March they return to their breeding-haunts, some nesting in inaccessible places at the " back of the Rock."

They commence about the 10th of March to build their nests, which resemble those of the House-Martin (*Chelidon urbica*). Placed in the roofs of caverns, these nests are very difficult to reach; and I did not succeed in examining the inside of one. The birds were sitting by the 30th of April. One locality for nests near Gibraltar, and the most accessible that I have seen, is a cave in a patch of rocks at the entrance of La Trocha, on the road from Algeciraz to Ojen, where it passes by the side of the ravine called la Garganta del Capitan.

At the back of the Rock, at Gibraltar, is a cave almost under the Osprey's eyry, which can only be entered by landing from a boat in fair weather. This cave is very large and open, with sand at the bottom sloping upwards for a considerable distance at a sharp angle.

The end of this cave, judging from the tracks of divers Genets or Striped Cats (*Viverra*), seems to be the regular dining-room of those animals; for whenever I have visited the place, it was covered with the tail-feathers and pinions of numbers of Rock-Martins mingled with those of a good many Swifts, Rock-Doves, and a few Lesser Kestrels.

Family LANIIDÆ.

127. LANIUS MERIDIONALIS, Temm. The Spanish Grey Shrike.

Spanish. Alcaudon real.

This Shrike is scarce in the neighbourhood of Gibraltar, the few specimens I have seen from there having occurred in autumn. Further north, though rather local, they become common in many places, being abundant and resident, breeding in the scrubby jungle near Seville. They nest, about the 15th of April, in bushes and low trees. The nest is large, the internal diameter measuring some five inches. When placed on a bough, the lower half is sometimes made of mud, the upper half being constructed with rough grass lined with fine grass, the whole covered outside with lichens and bits of some sort of *Scleranthus*—the same plant so much used by the Woodchat and many other birds. The eggs are from four to five in number. At one time I was under the erroneous impression that these birds were migratory, from seeing them near Gibraltar in autumn; but, never having seen or heard of them on the African side, I must have been mistaken: besides, they were more numerous about the Coto del Rey in winter than in May. However, this tends to show that they shift their ground in Spain, though not migrating out of it.

128. LANIUS ALGERIENSIS, Less. The Algerian Grey Shrike.
Moorish. Bou-seroond (*Favier*).

Favier's notes, which under the synonym "*meridionalis*" refer to this species, do not add any thing to the account I have written.

This Shrike, which would be more fitly named the North-west-African Grey Shrike, as it is not peculiar to Algeria, is not common in the immediate vicinity of Tangier; but a little further south, near Larache, and towards the Fondak on the road to Tetuan they are very abundant. Their habits, nests, and eggs are identical with those of *L. meridionalis*, their Spanish representative.

They are only met with in scrubby jungle. On the 18th of April I found one nest in a thick lentiscus bush, with five eggs, which were hard sat-on, and another on the 20th, with only three fresh eggs.

The species is distinguishable from the Spanish Grey Shrike by the grey colour of the underparts, by the stout bill and legs, and by the much larger alar speculum.

Length 9·6 inches; tarsus 1·1.

129. LANIUS AURICULATUS, P. L. S. Müll. The Woodchat Strike.
Moorish. Raicha el kra (*Favier*). *Spanish.* Alcaudon.

As common in Morocco as in Andalucia, the Woodchat arrives in March and April, leaving in August and September. The first arrival noticed at Gibraltar in 1868 was on the 3rd of April, in 1869 on the 3rd of April, in 1870 on the 29th of March; the passage ceases about the 20th of April. The latest seen was on the 14th of October 1871; in 1869 I observed them returning south on the 26th of August.

The Woodchat is one of the most abundant and conspicuous birds in spring on both sides of the Straits. Very tame and confiding, unlike their big cousins *L. meridionalis* and *L. algeriensis*, their pied appearance and the bright chestnut-coloured head of the adult males cause them to be noticed even by the most unobservant. They are to be seen in every direction in woods and on plains, perched on tops of trees, bushes, aloes, and tall plants, making their larders on the spikes of the aloes, and impaling on the thorns beetles, bees, and all kinds of insects. They are very mischievous among bees.

The nest, which usually contains eggs by the 10th of May, is a small edition of that of *L. meridionalis*, but is more covered outside with the greyish white flowers and stalks of the *Scleranthus*, and is usually placed low down in trees. The eggs, four or five in number, sometimes six, are subject to great variation, many resembling those of our English Butcherbird, *L. collurio*.

130. TELEPHONUS ERYTHROPTERUS (Shaw). Hooded Shrike.
Moorish. Abermat (*Favier*).

"This Shrike is nearly as common as *Lanius algeriensis* near Tangier, and is resident, nesting in bushes twice a year—in May or June, and again in November. Their note, which is a kind of whistle, harmonious and well-sustained, and very like that of a Blackbird, is usually heard from the middle of some thick bush (where they have a habit of hiding themselves), as well as in the thickest part of trees. They lay about three eggs, of the same shape as those of other Shrikes, but white marked with lines and small spots of ash-brown and russet, mostly at the thick end. The sexes are alike in plumage, and undergo no change."—*Favier*.

According to my limited experience, the Tchagra is rather scarce near Tangier, but more plentiful about a day's journey south. On the 25th of April I took a nest in a small tree close to the ground, containing three eggs slightly sat-on, which is, I am informed, the usual complement. The nest was not so compact as that of the Woodchat, containing less grass and dried flowers, being chiefly built of fibrous roots. It is an easy bird to recognize when it has been once seen on the wing, the chestnut of the wing-coverts being very conspicuous.

Favier states that they cross the Straits; all I can say is, I never saw a Spanish specimen, though a birdcatcher at Tarifa professed to recognize a skin and called it "Alcaudon carnicero," a name I had not heard before, but which I have since ascertained is applied to other Shrikes; so in any case, if ever it does occur in Spain, it is extremely rare, and as yet unobserved.

Family MOTACILLIDÆ.

131. MOTACILLA ALBA, Linn. The White Wagtail.

Moorish. Mizizi *(Favier).* *Spanish.* Pispita.

"This is the most abundant of the Wagtails, and passes the winter near Tangier, arriving during September and October, leaving in March. They are to be seen in large flocks following the plough, twittering incessantly."—*Favier.*

The above notes apply equally to the White Wagtail in Andalucia; but it was never seen by me after the 16th of March. They roost together in great numbers on the short rushes and grass in the marshes at Casa Vieja, where also I have often seen them on horses' backs, picking off vermin and catching flies, doing the work of *Ardeola russata.* Length $7\frac{3}{8}$ to 7 inches.

132. MOTACILLA YARRELLII, Gould. The Pied Wagtail.

"This is the most rare of the Wagtails near Tangier; they are found at the same times and places as *M. alba.*"—*Favier.*

I obtained a specimen in summer plumage at Tangier; it is scarce enough in Andalucia in that plumage, but probably in winter dress escapes unnoticed. Length $7\frac{1}{2}$ to 7 inches.

133. MOTACILLA SULFUREA, Bechst. The Grey Wagtail.

This Wagtail is stated by Favier to be a common winter

visitor near Tangier, appearing in September and October, departing in February and March.

In Andalucia they are most abundant on passage and during the winter months, but many pairs nest along the mountain-streams of the sierras—three or four pair particularly on the river which runs down the Garganta del Aguila, the valley in which is the "waterfall" of Algeciraz. They build usually in holes of the brickwork of the water-mills, sometimes close to the wheel. A pair also regularly nest at the mill in the Cork-wood; and Mr. Stark, when with me on the 9th of April, found a nest built in a hole of a large rock overhanging the Guadalmalcil, a mountain-torrent between Tarifa and Algeciraz. This nest was placed out of reach of either man or beast; the hen bird was visible from the opposite side, and apparently sitting hard. On the 19th of April, the nest at the mill in the Cork-wood contained one egg, while another nest I found on the 24th of May had four fresh eggs in it. This tends to show that they breed more than once in the season. The nest is constructed of grass and small roots lined with hair.

Some only of the males had the black throat on the 8th of March; but all had assumed it by the 1st of April. The females do not always exhibit this mark, some not having it at all; and in none is it so well defined as in the males. It is the largest of the Wagtails found in these parts, being about $7\frac{3}{4}$ inches in length, with the tail-feathers much longer in proportion to the body than in the other species.

134. BUDYTES FLAVUS, *vel* CINEREOCAPILLUS, Savi. The Grey-headed Yellow Wagtail.

This species is found on both sides of the Straits in great abundance; the earliest that I saw it was on the 20th and 24th of February (in different years), many appearing on the 25th. From that time till the 20th of April they continued to pass; and on that date I saw great numbers at Gibraltar, resting on the "flats" at Europa after their flight across the sea. They leave in August and September. They keep to marshes, nesting in grass and herbage at the edge of water, sometimes among the sedges, and lay in the end of April. The eggs much resemble those of *Calandrella brachydactyla*. *Budytes rayi*, the English form of Yellow Wagtail, I never met with on either side of the Straits.

135. ANTHUS SPINOLETTA, Linn. The Water-Pipit.

This Pipit breeds on the summit of the Sierra del Niño, though not in any numbers. One which I shot on the 24th of May was, from the state of the breast, evidently sitting. I had not time to stop to find a nest; and they were very wary. In winter they descend to the mud flats and sea-shore and are not uncommon. The specimen killed on the 24th of May had the basal half of the under mandible pink flesh-colour, the remainder of the beak being dark brown; the legs were whitish flesh-colour, the feet dark brown, the claws still darker; irides hazel.

136. ANTHUS OBSCURUS, Gm. The Rock-Pipit.

Favier states that this species occurs in winter near Tangier, and may be always seen on the sea-shore. I think (not having seen any of his specimens) that these remarks apply to *A. spinoletta*, which he does not mention in his MS.

My only reason for including the present species is, that I brought home a specimen shot among many others on the mud at Palmones, near Algeciraz, in March 1870, which was identified by Mr. Sharpe as being *A. obscurus*. Not having obtained any since, it is quite possible there may have been some error about it, and that the species does not occur in Andalucia.

137. ANTHUS PRATENSIS, Linn. The Meadow-Pipit.

Favier having confounded this species with the next, I have omitted his notes.

The Meadow-Pipit is equally common in Morocco and Andalucia from October to the end of March.

Length 5·8 inches; tarsus 0·9.

138. ANTHUS CERVINUS, Pall. The Red-throated Pipit.

I obtained two Pipits in 1870 on the 10th of March, which

I took home to England; and they were identified as belonging to this species by Mr. Sharpe and Captain Shelley. In 1874, on the 8th of March, I shot among a lot of *A. pratensis* another bird, which appears to be *A. cervinus*; but as the rufous throat is not developed, though it shows signs of that mark of the breeding-plumage, to determine with certainty to which species it really does belong is impossible; so I mention and include it in my list with a view to some future collector paying attention to the subject. For my own part I have little doubt that *A. cervinus* does occur on passage in Andalucia and Morocco.

139. ANTHUS TRIVIALIS, Linn. The Tree-Pipit.

According to Favier this Pipit is common near Tangier during migration, crossing the Straits in March and April, returning in October and November. On the Spanish side it is found on passage only: the 9th of April is the earliest date on which I noticed it; but I saw many about the 20th.

Length $6\frac{1}{2}$ inches; hind claw much curved.

140. ANTHUS CAMPESTRIS, L. The Tawny Pipit.

Moorish. Solist (*Favier*).

"Found near Tangier on passage in April, returning in August, but is not very common. They migrate in pairs and keep close together, so that it is very easy to shoot both at one shot."—*Favier.*

The earliest date on which I saw one at Tangier was on the 31st of March. On the Spanish side they appear to frequent high ground, as on the 1st of May I saw many on the open spaces about Ojen, and thence all along to the Venta de Subalbarro; they were not then nesting. I never met with them on the low ground.

141. ANTHUS RICHARDI, Vieill. Richard's Pipit.

I shot one on the 1st of March on the shore, evidently just arrived, and obtained others on the 20th of April, 1870, not again noticing it.

Length $7\frac{3}{4}$ inches; wing, carpus to tip 3·6 inches; the hind claw is very long and slightly curved.

Family ALAUDIDÆ.

142. ALAUDA ARVENSIS, Linn. Sky-Lark.

Spanish. Zurriaga, Terrera.

"Found near Tangier during winter, arriving in October and November, departing in March. They are found in immense flocks during their stay."—*Favier*.

The same is to be said of the Sky-Lark in Andalucia, where in some localities quantities are caught at night with a bell and a lantern. I have known a boy bring in six or seven dozen birds at a time. Calandras, Buntings, Larks, in fact any birds that sleep on the ground can be thus taken.

143. GALERITA MACRORHYNCHA, Tristram. Tristram's Lark.

This bird is very similar to *G. cristata*, but is larger, measuring about 7·8 inches in length, and having a longer, thicker, and more curved bill. It is mentioned by Mr. Drake as having been met with near the city of Morocco; I did not observe it near Tangier, and chiefly include it with a view to draw the attention of any ornithologist visiting the Straits to the subject.

144. GALERITA CRISTATA (Linn.). The Crested Lark.

Spanish. Carretera.

The Crested Lark is one of the most abundant birds both in Morocco in Andalucia, though never seen in any great numbers together. They are distributed in pairs on every road-track and open plain, often at intervals of only some twenty yards. Excessively tame and fearless, they have acquired the name of *Carretera*, from their habit of frequenting roads, to which they resort as much on account of the horse- and mule-dung, at which they are to be seen pecking, as for the purpose of dusting themselves; and they are often to be noticed on the sea-shore, running about like a Sanderling within a yard of the water. They have no song worthy of the name, and are altogether rather vulgar and uninteresting birds. This species is one of those which I could not detect migrating at all.

The Crested Lark usually commences to lay about the 20th of April, placing its nest in some tuft of grass or under shelter of a small stone or clod of earth—constructing it, like those of other Larks, with bits of grass bents &c., lined with hair.

One nest which I found was placed between the tracks of a much frequented road near Tangier, in such a position that every passing animal must have touched the small clumps of grass under cover of which the nest was built. Now, was this site chosen because snakes, lizards, and other vermin would be less likely to come on the beaten track? I cannot help thinking that birds in many instances have instinct enough to breed close to houses and roads with a view to obtain protection from their enemies through the presence of man.

The Crested Lark is subject to great difference in the depth of the tints of the plumage. This difference of colour varies according to the soil, and has been very puzzling to cabinet naturalists, as the enormous list of synonyms will testify. Mr. Dresser, in his article on this bird, gives no less than thirty-seven different names, which have for the greater part been manufactured on account of the variations of plumage and size noticed in this species.

145. LULLULA ARBOREA (Linn.). The Wood-Lark.

Spanish. Alondra de monte.

According to Favier, this species "occurs near Tangier on passage during March."

On the Andalucian side the Wood-Lark is sparingly and locally distributed during the winter months up to as late as the 21st of April, frequenting scrub where not very thick, a favourite locality near Gibraltar being the Chapparales in the Cork-wood. Well known to the Spanish birdcatchers, and highly valued as a cage-bird; I am assured by them that the Wood-Lark never remains to nest near Gibraltar.

This species is rather short-tailed, and has a light-brown eyebrow.

146. CALANDRELLA BRACHYDACTYLA (Leisl.). The Short-toed Lark.

Spanish. Terrera.

"This bird is found on passage near Tangier, crossing in very large flights during March and April, returning in August and September. Many remain to breed, frequenting the same localities as the Calandra Lark."—*Favier.*

On the Andalucian side of the Straits, my experience of the Short-toed Lark is that it chiefly migrates in pairs during the spring; and I never met with any of the above-mentioned large flights. The spring arrival commences about the middle of March; and the passage continues for a month later, at which time nests with eggs may be found near Gibraltar. Excessively abundant, as above stated, in the same situations as the Calandra; they prefer fallow ground, nesting under shelter of some clod or in any slight depression of the ground. I never could find the nest myself, except by putting the old bird off. A very good way of finding the nests of all the Larks and ground-breeding birds is, with the assistance of a man to hold it at the other end, to drag a rope about a hundred yards long across the ground, being careful that it drags *on* the ground; directly a bird flies up, drop the rope, go to the spot, and in all probability a nest will be found.

147. CALANDRELLA BÆTICA, Dresser. The Andalucian Short-toed Lark.

This species was found by Lord Lilford in the corn-land on the banks of the Guadalquivir below Seville, where it is known to occur from February to the summer. Whether it is migratory is not yet known; but it is almost certain to be so.

Differs from *C. brachydactyla* in having the breast and back marked with blackish brown: it is very like *C. minor*; but the same markings are more defined than in that species.

148. MELANOCORYPHA CALANDRA (Linn.). The Calandra Lark.

Spanish. Alondra, Calandria.

The Calandra is extremely abundant and resident on both

sides of the Straits, gathering together in flocks during the winter.

Frequenting all the vegas or plains in Andalucia, it is, from its numbers, size, note, and peculiar varying flight, very conspicuous; in some localities it positively swarms. At times their flight is very like that of some of the smaller Waders; and often when flying in the breeding-season they utter notes which very much resemble the cry of the Green Sandpiper, only of course not so loud.

They consort much with the Short-toed Lark, and nest sometimes close together on cultivated as well as pasture land, laying about the second week in April.

It is a very common cage-bird at Gibraltar, and much prized for its song as well as for its lively habits.

149. OTOCORYS BILOPHA, Temm. The Horned Desert-Lark.

This species is mentioned by Mr. Drake as having been obtained in Morocco; but I never had the good fortune to meet with it. The upper part of the body is a pale fawn-colour, the adult male having a black gorget and two long black ear-tufts.

Family EMBERIZIDÆ.

150. EMBERIZA MILIARIA, Linn. The Common Bunting.

Moorish. Dorris (*Favier*). *Spanish.* Triguero.

This well-known bird is exceedingly numerous both in Andalucia and Morocco, and, being to a great extent migratory, is perhaps least plentiful in winter. From its fearless stupidity and from its conspicuous habit of perching on the top of some small bush or plant, vast quantities are killed by the Spaniards and exposed for sale in the markets, while during the winter months a great many are caught roosting on the ground by the aid of the lantern.

The common Bunting commences to lay about the first week in May, often placing its nest at the edge of marshes; and I have taken it in the midst of a swamp, placed on a dry tussock, within a yard of a nest of Savi's Warbler. Albino and light-coloured varieties of this Bunting are very frequent in Spain as well as in other countries.

151. EMBERIZA CIRLUS, Linn. The Cirl Bunting.

Spanish. Linacero.

According to Favier's notes, this Bunting is as common around Tangier as it is near Gibraltar, migrating northwards in March, returning in October and November, remaining during the winter months; many also stay to breed during April.

The Cirl Bunting is very frequent all along the coast from Algeciraz to Tarifa, and from its tameness, and the bright colour of the males, with their conspicuous black throats, is sure to attract attention. It appears to be more of a tree-frequenting bird than the Yellowhammer (*Emberiza citrinella*), and is especially fond of the glades and open patches of cultivation in the Cork-wood.

The least wing-coverts of both sexes, in the young and old birds, are *olive-green.* This will serve to distinguish them from *E. citrinella,* which species I have never seen or heard on either side of the Straits.

152. EMBERIZA HORTULANA, Linn. The Ortolan Bunting.

Moorish. Merskezan (*Favier*). *Spanish.* Hortolano.

"The Ortolan is, next to the Common Bunting, the most abundant of the genus near Tangier. Some remain to breed; while the rest pass on during April, returning in September. Migrating in large flocks, they prefer wet ground, and are not observed in the winter months."—*Favier.*

Curiously enough, I never succeeded in obtaining the Ortolan nearer to Gibraltar than in the vicinity of Casa Vieja, where I shot it in May. In the vicinity of Seville they are plentifully met with, but do not occur during the winter.

153. EMBERIZA CIA, Linn. The Rock-Bunting.

Although not mentioned by Favier as found in Morocco,

I have seen specimens obtained there, and met with a pair in April near Jebel Moosa.

On the Spanish side of the Straits it is a common and, like most of the Buntings, a stupidly tame bird, as far as my experience goes, living about stony, rocky, and hilly ground. Till 1874 I never noticed them perching on trees; but in the spring of that year I saw three different birds, when disturbed, settle on trees and bushes. At Gibraltar they are met with in winter, but disappear in the spring. I have shot them at the back of the Rock when looking for Alpine Accentors, in company with which birds I have seen them feeding on the refuse-heap at the signal-station. In April they keep to the slopes and tops of the sierras, nesting during that month.

154. EMBERIZA SCHŒNICLUS, Linn. The Reed-Bunting.

Included by Favier in his list as "rare near Tangier; met with in December." I imagine that he considered it rare from not observing the females or young birds, which on the Andalucian side are, during the winter months, very plentiful in wet suitable localities; the more conspicuous, old, black-headed males, with the ringed neck, are not so common. They are most abundant near Gibraltar from December to February; and I have seen them on passage as late as the 7th of April. They do not remain to nest in the sotos at Casa Vieja; but near Seville, where they are often sold in cages under the name of *Hortolano*, I have seen them in May, and have no doubt that they there remain throughout the breeding-season.

155. PLECTROPHANES NIVALIS (Linn.). The Snow-Bunting.

The Snow-Bunting has only been recorded once from the Moorish side of the Straits; and this occasion was mentioned by Mr. Drake in his list of the birds of Morocco, published in 'The Ibis' for 1867 (p. 427). This specimen, lately in the possession of Olcese (Favier's successor at Tangier), was a female, and in fine plumage.

I never heard of an instance of the occurrence of the Snow-Bunting in Andalucia.

Family FRINGILLIDÆ.

156. FRINGILLA CŒLEBS. The Common Chaffinch.

Spanish. Pinzon.

Our well-known English Chaffinch is throughout Andalucia as common as it is in England where there are trees—most frequent in winter, the residents being then greatly reinforced by those which arrive from the north. It appears on the rock of Gibraltar as early as the 12th of August; but I do not think it ever remains to nest. Great numbers breed in the Cork-wood and in all the wooded valleys of the neighbourhood, usually commencing to lay about the end of April.

I am not aware of the occurrence of this Chaffinch on the African side of the Straits, but cannot help suspecting that stragglers must occasionally cross.

157. FRINGILLA SPODIOGENA, Bp. The North-African Chaffinch.

Moorish. Bourdo.

Favier does not appear to have distinguished the difference between this and the European bird (*Fringilla cœlebs*), though Mr. Drake mentions it. It is the Chaffinch of the country, and is in Morocco, about Tangier, Tetuan, &c., as common as its congener is in Andalucia, and has the same habits, the eggs and nest being also like those of *F. cœlebs*; but the note, harsher and not so musical, is very distinct. It does not appear ever to cross to the European side of the Straits. The adult male is to be distinguished from *F. cœlebs* by the back being *yellowish green,* and by the throat and breast being light yellowish buff. The females are similar to those of the common Chaffinch.

158. FRINGILLA MONTIFRINGILLA, Linn. The Brambling or Mountain-Finch.

Spanish. Montañes, Millero.

This species "has been once obtained near Tangier, in 1845,

when I killed a female from among a lot of Linnets; the male escaped."—*Favier.*

The Brambling, on the Spanish side of the Straits, is of very irregular though not unfrequent appearance near Gibraltar; in the winter of 1870–71 they were common, as well as Siskins. Near Seville they are more regular in their appearance, and are sufficiently well known about Cordova to bear the local name of "Millero."

159. FRINGILLA NIVALIS. The Snow-Finch.

I have seen specimens of this bird obtained in the Sierra Nevada, but never observed it myself; indeed, its habitat being confined to Alpine ranges, it is a bird little likely to be noticed or obtained.

160. PASSER MONTANUS, Linn. The Tree-Sparrow.

This Sparrow occurs sometimes in Andalucia, as I have seen specimens obtained in the country; but I was not fortunate enough to fall in with it personally. The adult males are to be distinguished from those of the House-Sparrow by their smaller size, the top of the head being *chestnut*-coloured, with a black spot on each side of the neck.

161. PASSER DOMESTICUS, Linn. The Common Sparrow.

Moorish. Bertal. *Spanish.* Gorrion.

Common on both sides of the Straits, being the Sparrow of the district.

The Italian Sparrow (*Passer cisalpinus*) I included in my list of birds of the south of Spain, from having seen a specimen said to have been obtained at Seville; but on second thoughts, not having seen it in the flesh, I consider it better omitted, especially as its occurrence so far west amongst thousands of *P. domesticus* would be most improbable. Mr. Gould, however, showed me a specimen obtained "somewhere in Spain" (rather a large district!) by the late Captain Cooke Widdrington.

This species only differs from *P. domesticus* in the adult male having the *top* of the head and nape of the neck chestnut, the cheeks purer white, and the eyebrows white. The females are not to be distinguished from those of that species.

162. PASSER SALICICOLUS, Vieillot. The Spanish Sparrow.

This is another of the chestnut-headed Sparrows, and is local in distribution on both sides of the Straits. It is in some places very abundant; and, as is well known, they often build under the nests of the larger birds of prey. I found one nest built underneath a nest of *Buteo desertorum* in April.

The females resemble those of the common Sparrow. The adult males have the upper part of the plumage much marked with black.

163. PETRONIA STULTA (Scop.). The Rock-Sparrow.

Spanish. Gorrion montes.

Neither Favier nor Mr. Drake mentions having seen this Sparrow in Morocco, where, however, it is found, as on the Spanish side, commonly in the sierras and rocky ground, nesting in May in holes of rocks.

The adult male has a yellow crescent on the throat; in the female this mark is much fainter.

164. CHLOROSPIZA CHLORIS, Linn. The Greenfinch.

Spanish. Verdon.

"Found near Tangier as a common resident; others migrate in immense flights, which pass north in February and March, returning in October and November."—*Favier*.

This species, another of our common British birds, is extremely abundant on both sides of the Straits. Many are resident, nesting during the month of May; and hundreds are caught in August and September and brought into the markets, where they are exposed for sale in large bunches. The Greenfinch is also a very common cage-bird; for sometimes I have seen as many as twenty, each in a separate cage, hanging outside the wall of a house. What its merits as a song-bird are, I never could understand, as its song is to me positively unpleasant.

Greenfinches from Morocco and the south of Spain are rather smaller and more brightly coloured than English birds, and have been supposed to belong to the form called *Chlorospiza chlorotica* (Licht.); but I never could see sufficient reason to separate them as a species.

165. LINOTA CANNABINA, Linn. The Common Linnet.

Moorish. Sharif (*Favier*). *Spanish.* Camacho, Jamas.

"Abundant around Tangier, many being resident and nesting from March to June. They are mostly migratory and cross to Europe in March and April, returning in large flocks during September and October."—*Favier.*

The common Linnet is very plentiful on the Spanish side, especially during the winter months. Great numbers remain to breed, nesting in April, mostly in scrub on the sides of the hills. Upon one occasion a pair built on an olive-bush in my garden at Gibraltar.

The adult males are, as a rule, far more brightly coloured than the average of English specimens.

166. LINOTA RUFESCENS (Vieill.). The Lesser Redpole.

This bird is not mentioned, but is included by Mr. Drake in his list of the birds of Morocco.

On the Spanish side of the Straits it can only be considered a very rare and (like the Twite, the Siskin, and the Brambling) a very irregular winter visitant.

167. LINOTA MONTIUM (Gm.). The Twite, or Mountain-Linnet.

I have no record of the occurrence of this Linnet on the Moorish side. In Andalucia it is as rare a winter straggler as the Lesser Redpole.

The Twite has no red colour on the head or breast; the tail is much forked; and in the adults the bill is yellow.

Length 5·25 inches; wing, from carpus to tip, 3 inches.

168. CARDUELIS ELEGANS, Steph. The Goldfinch.

Moorish. Mouknin. *Spanish.* Gilguero. *Provincial.* Silguero.

"Exceedingly plentiful near Tangier, and resident; but numbers migrate, arriving from about the month of August, to depart again for the north in the month of March."—*Favier.*

The Goldfinch is, without doubt, the most common and abundant bird in the west of Andalucia. Always plentiful in every direction, they appear in countless flocks when the seed of the various thistles becomes ripe; and Spain is the country *par excellence* both of thistles and donkeys. The former, in some of the vegas and plains, grow in regular impenetrable thickets, in places covering acres of ground; for when the land is left fallow for a season, all weeds are allowed to run riot, and they do so with a vengeance. Some of the thistles (and there are many different kinds) are very handsome—a large, yellow, carline species being perhaps the most attractive to the eye. The stalks, heads, and leaves of another sort, very like the garden artichoke, are a staple vegetable with the Spaniards, who sometimes were disposed to be indignant when I remarked, in fun, that in England donkeys, not men, eat thistles (or *cardo*); but at the same time I omitted to mention that we grow artichokes in our gardens. The stalk of another thistle is much used as tinder in the rural districts, and known as *yesca de cardo*; it takes light well from the sparks made by a flint and steel, most of the peasants using no other method of lighting their *papelitos.*

To return to the Goldfinches, at the time of their thistle-harvest they are caught in vast quantities in clap-nets; and it is not unusual to see a man with bunches of several hundreds, which are sold at a ridiculously low price. Though perhaps rather dry, they are not to be despised as morsels; but one feels as if committing a grievous sin when devouring such a charming little bird.

The Goldfinch in Andalucia breeds about the beginning of May, and occasionally nests at Gibraltar in the Alameda and various gardens at the South.

169. Chrysomitris spinus, Linn. The Siskin.

Spanish. Lūgano.

I have not any information of the occurrence of the Siskin on the African side of the Straits. In Andalucia they are very irregular in appearance, in some winters not being noticed at all. The Spaniards say they only come every seventh year. This idea is prevalent about Seville, as well as near Gibraltar; but is, I need not say, a popular error.

In the winter of 1870-71 they were plentiful wherever there were any alder trees; and I saw some as late as the 4th of April. In the two previous winters, and during the one following, none were obtained by the birdcatchers, who are always looking out for them, as they are much desired and fetch a good price as cage-birds. During my last visit I saw four on a cotton-poplar tree in the Alameda at Gibraltar, on the 24th of March; they were so tame as to allow of my approach within a yard of them, and remained for a long time close to me.

170. Serinus hortulanus, Koch. The Serin Finch.

"This bird is very abundant near Tangier, both as a resident and on migration, when they are seen passing north in immense flights during February, returning in October and November."—*Favier*.

The Serin Finch is found on the Spanish side in accordance with the above note. During the breeding-season they greatly frequent the Cork-wood, and their hissing unpleasant song is to be heard all around. They chiefly seem to keep to the banks of rivers, nesting in May on trees and bushes, like the Goldfinch, resembling that bird both in their nest and eggs.

The Citril Finch (*Dryospiza citrinella*) I never succeeded in obtaining on either side of the Straits.

171. Carpodacus githagineus, Licht. The Desert-Bullfinch.

This species is mentioned by Mr. Drake as being seen in the south of Morocco.

The adults are easily known by their rosy-tinted plumage

and red bill. In the young these colours do not exist, but the bill is pale yellowish brown, and the plumage sandy-coloured.

Length 5 inches, tarsus 0·7*.

172. LOXIA CURVIROSTRIS, Linn. The Common Crossbill.

Favier only mentions having obtained this bird once near Tangier, "a specimen being picked up in a dying state in 1855."

Although never having myself met with this Crossbill on either side of the Straits, I have seen undoubted Andalucian specimens. I regret to be unable to mention when or where it is to be looked for.

173. COCCOTHRAUSTES VULGARIS, Brisson. The Common Hawfinch.

Spanish. Cascanueces (Nutcracker).

Favier states the Hawfinch to be " very rare near Tangier, having only met with two—one in 1836, the other in 1849."

On the Spanish side of the Straits this bird is very common, and most plentiful in winter. Some nest in the Cork-wood in May; and during the season of migration, they often frequent pine-woods, and are then rather shy and difficult to approach. About Cordova they are most abundant, and are there and at Seville exposed alive for sale at about one real apiece. I kept a pair, which I purchased at Seville, for some time; but never could tame them. The hen bird at last killed her mate, having previously at regular intervals plucked him while living.

I gave this amiable and domestic female to a bird-fancier at Gibraltar, much to his delight, but ultimately to his sorrow, as she vented her temper upon some other pet birds with which she was caged, and, in consequence, justly suffered capital punishment.

* Since writing the above, an immature specimen of *Carpodacus erythrinus*, the Scarlet Grosbeak, has been obtained by Mr. H. Saunders from a collection at Malaga. The females and young males of this species are at first sight very likely to be mistaken for immature Greenfinches (*C. chloris*), but are to be distinguished by the form of the bill and the two distinct yellowish white bars on the wing. In this plumage it is doubtless the *Fringilla incerta* of Risso. Total length about 5·8 inches.

Family ORIOLIDÆ.

174. ORIOLUS GALBULA, Linn. The Golden Oriole.

Moorish. Tair es sfar (Yellow bird).   *Spanish.* Oropendola.

According to Favier the Golden Oriole " crosses the Straits in great numbers during April and May, returning in July, August, and September." These dates much agree with my own observations on the Spanish side, I having first seen them in 1869 on the 21st, in 1870 on the 18th, in 1871 on the 4th (one only), and in 1872 on the 11th of April: many passed on the 16th in that year. The spring migration lasts up to the 14th or 15th of May. Some few pairs remain to breed in the vicinity of Gibraltar; but the majority pass further north and resort to fruit-producing districts, where they get the credit of doing much damage to cherries, mulberries, &c.

Almost entirely a fruit-eating bird, those who have kept them alive informed me that they could not preserve them through the winter—nor, indeed, longer than fruit was to be obtained.

When the loquats were ripe in my garden at Gibraltar in May 1870, the male Golden Orioles remained about as long as the loquats lasted, but would not admit of much observation, as they are very shy and difficult to watch. They are more often heard than seen; and I have spent hours in trying to get a shot as they skulked in the thickest foliage of tall trees, continually piping their flute-like note.

Some are always to be heard during May near the Mill and the "Second Venta" in the Cork-wood; and a pair usually frequented the lower part of the First Pine-wood. I found one nest in the middle of May, built at the very extremity of a bough at the top of a high oak tree; but it was impossible to obtain it without cutting the branch off.

The young male Golden Oriole resembles the females in its more sombre greenish plumage.

Family CORVIDÆ.

175. CORVUS CORAX, Linn. The Common Raven.

Spanish. Cuervo.

The Raven is found sparingly but very generally distributed on the Spanish side of the Straits, but does not seem to be found on the Moorish side. They are resident, and commence to lay about the middle of March, thus, very curiously, breeding later in Andalucia than in England (or, I had better say, in the north). One pair nest at Gibraltar, and, as is customary with Ravens, are the terror of all birds that approach their domain. Another pair nest at Casa Vieja, in the old quarry called la Cima, just outside the village.

In no case that I have seen have their nests been in any thing like proximity to one another, the reverse being the case with *Corvus tingitanus*, that bird not showing such jealousy of its brethren.

176. CORVUS TINGITANUS, Irby. The Tangier Raven.

Moorish. Grâb.

This species or race appears to me to be quite distinct from *C. corax*, and was noticed and described by me in 'The Ibis,' for 1874 (p. 264).

Smaller than the Common Raven, *C. corax*, it differs in the shape of the bill, which most resembles that of *C. culminatus* of India. The note is different from that of *C. corax*; and its very gregarious habits are opposed to those of our common Raven.

Many specimens are very much marked with rusty brown on the wings and tail, others very slightly so. In all that I have seen there is a tinge of brown on the wings—not that this coloration is of any consequence in determining it as a distinct species. They also breed later than *C. corax* does on the Spanish side.

This Raven is exceedingly abundant around Tangier and along the coast as far as some distance south of Larache. I did not observe any in the high parts about Apes' Hill. Outside Tangier, flocks of them may be seen feeding on the refuse which is carried from the town and thrown on the sea-shore. They are exceedingly tame to the natives, being viewed with superstitious awe by the Moors, but are wide awake to the European, especially if he carries a gun, and if once fired at will not give a second chance. The only way to be sure of getting them is at the nest, which is constructed of sticks, neatly lined with grass and small roots, and is built in clefts of rocks, on trees and in low bushes; one nest which I saw was fixed in the crook or angle formed by a dead flowering stalk of the aloe, which had fallen across another stalk in full flower.

The eggs are usually laid about the 20th of April, and vary in number from five to seven, and, like those of others of the Crow tribe, differ much in the markings. Favier in his MS. says of this Raven, under the head of *C. corax*, " This species is another of those birds for which the Mahometans evince a superstitious feeling, the liver, tongue, brain, and heart, of the Raven being considered antidotes against the effects of the evil one; the same virtues are attributed to the feathers and heart of the Hoopoe. The Raven is the only species of Crow found in the neighbourhood of Tangier, and is very abundant."

Length 18·5 inches, wing 14·5, tail 8, tarsus 2·5, bill from gape, 2·5.

177. CORVUS FRUGILEGUS, Linn. The Rook.

I never met with the common Rook near Gibraltar, or, indeed, further south than the Coto del Rey, in the neighbourhood of Seville, where there were several large flocks in January; and it appears to be there a regular winter visitant.

On the Moorish side of the Straits I can find no record of its occurrence.

178. CORVUS CORONE, Linn. The Carrion-Crow.

Not mentioned by Favier, but is included by Mr. Drake in his list of the birds of Morocco; I never met with it on the African side. On the Spanish side it is scarce, and I only remember one nest, the date of which I have mislaid; but it was taken near Utrera during the month of March.

179. CORVUS CORNIX, Linn. The Hooded Crow.

Not recorded from the Moorish side of the Straits. Mr. Saunders mentions the Hooded Crow as having been met with in Andalucia; but I never fell in with one or saw an authentic specimen. I noticed one in a museum at Seville; but upon inquiry it proved to be from France; so, although I have included it in my list, it is necessary to obtain further information before considering it even a rare straggler in Andalucia.

180. CORVUS MONEDULA, Linn. The Jackdaw.

Mr. Drake mentions having met with the Jackdaw near Tetuan, where I did not see it; nor, indeed, did I find it anywhere on the African side. Favier also omits the bird from his list. Probably, if it occurs, it is, as on the Spanish side of the Straits, extremely local, the only locality in which I have seen any being the Coto del Rey, near Seville, where, in 1870, it was abundant, nesting about the end of April in holes of trees, one or two pairs utilizing the roof of the Palacio.

181. PYRRHOCORAX GRACULUS, Linn. The Common Chough.
Moorish. Narrar. *Spanish.* Graja.

This species is stated by Favier to be found in large flights near Tetuan—a statement I can fully corroborate. I also saw a great many about the cliffs of Abyla, or Apes' Hill, opposite to Gibraltar.

In the rocky sierras of Andalucia the common Chough is plentiful, particularly about Ubrique. They are, of course, resident; but I am unable to state the time of nidification. In this species the bill is the same colour as the legs, viz. vermilion.

182. PYRRHOCORAX ALPINUS, Vieill. The Alpine Chough.

I have no record of the occurrence of this Chough on the African side, though I imagine that I saw it at Apes' Hill, where a bird flew so close to me that I thought I could distinguish the yellow bill. On the Spanish side it occurs near Granada.

The yellow bill, shorter than that of the common Chough, will serve to distinguish this species. The legs in both are the same vermilion colour; but the present bird is said to inhabit higher ground.

183. PICA RUSTICA, Scop.   The Common Magpie.

Spanish. Urraca, Marica.

Our Common Magpie is extremely local in Andalucia; but where met with it is very abundant. It does not, however, occur to the south of Seville, except on the banks of the Guadalquivir to below Coria, just as far as there are any trees and bushes. Hundreds frequent the Coto del Rey, where they breed in the beginning of May, accommodating the Great Spotted Cuckoo with their nests.

The Spanish race undoubtedly runs into the African form *P. mauritanica.*

184. PICA MAURITANICA, Malh.   North-African Magpie.

This species, which, however, I failed to meet with in Morocco, is the Magpie of the country; perhaps it is very local, as Mr. Drake describes it as abundant in the parts he visited. Is distinguishable from *P. rustica* by the bare space behind the eye and by the black rump, both species being identical in size.

The Nutcracker, *Nucifraga caryocatactes,* I never met with on either side of the Straits, though I saw one in a collection at Cordova; but from what locality was not stated; so, until some further evidence be obtained, it cannot with certainty be included as an Andalucian bird. It has been recorded from Estremadura by Captain Cook Widdrington, and possibly may be found in some of the high wooded ranges.

185. CYANOPICA COOKII, Bp.   The Spanish or Azure-winged Magpie.

Spanish. Mohino rabilargo.

This species is peculiar to the Peninsula, but does not occur in the vicinity of Gibraltar. The nearest locality to that place where it is to be found is about Coria del Rio, below Seville; thence, as far as the Coto del Rey, it occurs in tolerable numbers, but is much more common towards Cordova, and, as I am informed, swarms in some parts of Estremadura. It is, however, a very local bird. It appears to me not to have much of the habits of the true Magpies, but some of those of the Jays; but my acquaintance with it is very limited.

The nests which I have seen were built on boughs at no great height from the ground, rather clumsily constructed with small sticks, grass, moss, and wool—containing five eggs; but as many as seven are frequently found. They are well figured in 'The Ibis' (1866, p. 382, pl. x. figs. 3–8), from specimens obtained by Lord Lilford in 1864, and vary a great deal in colour and markings, the commonest form being of a stony buff colour marked with purplish and brown spots. I kept four of these birds, reared from the nest, for some time alive, feeding them on grapes, figs, bread, beetles, and grasshoppers. Always placing the insects under their feet, they picked them to pieces much as a Hawk or an Owl tears its prey. They became very tame and amusing; but during my temporary absence, unluckily, all died, to my great disappointment, as I meditated bringing them home to England. I never heard of this bird on the Moorish side of the Straits.

The sexes are alike in plumage.

Family STURNIDÆ.

186. STURNUS VULGARIS, Linn. The Common Starling.

Moorish. Zarzor. *Spanish.* Estornino.

" This bird arrives about Tangier in large flights from October to January, departing in March. During the autumnal migration the flights are often mixed up with *S. unicolor*. In October, 1842, a Moor brought to Tangier about three hundred and fifty Starlings, which he affirmed he had caught at one time in a net; about half of these birds were *S. unicolor*."
—*Favier*.

With all due respect for the memory of M. Favier, I cannot but suspect that the above-mentioned birds were in reality all common Starlings, half adult and half immature birds, as I never saw the two species migrating or consorting together.

In Andalucia, the common Starling may be said, roughly speaking, to come and go with the Golden Plover. The earliest date on which I noticed their arrival was the 15th of October, the latest date on which I saw any being the 1st of March. I have a note of seeing many thousands passing southwards in successive flights on the 28th of October. During the winter months they are seen in swarms about low ground; and the Spaniards shoot immense numbers at their roosting-places in the reed-beds near Veger and Casa Vieja. Consequently, during their stay, Starlings form a very cheap and, I may fairly say, nasty dish in all the ventas and ventorillas in the vicinity.

187. STURNUS UNICOLOR, Marm. The Sardinian Starling.
Moorish. Zarzor kahal (Black Starling). *Spanish.* Tordo.

"This Starling is very abundant around Tangier, passing north in March and returning during the month of September, many, however, remaining to breed."—*Favier.*

The Sardinian Starling, as it has been termed (Spotless Starling would be, perhaps, a more appropriate and distinctive name), is almost entirely migratory in Andalucia; but I have seen them there in December. Not so abundant as the common Starling, they resemble that bird in their habits and note, nesting about the end of April in roofs of houses in towns, and they make much use of the old Moorish towers, besides building in holes of trees; the eggs exactly resemble those of *S. vulgaris*. They are more common some sixty miles north of Gibraltar than in the immediate vicinity. Three or four pairs used to frequent the Venta at Casa Vieja, and during November and December, when I was there, nearly every morning assembled on the roof, whistling and pluming themselves before going forth for the day. The *amo*, or landlord, well known as "old Bernardo," begged me not to kill them—a request I most scrupulously complied with; but on my return there in 1874, they were absent, probably killed by some of the shooting visitors from Gibraltar. Whether the death of the old man caused them to lack protection I cannot say. This old fellow, who had served as a sergeant in the Spanish army, and was present at the defence of Tarifa in 1811, was a fine specimen of the Spaniard, and used to tell wonderful stories of his soldiering days. I regret that since his decease the Venta has changed for the worse, both in prices and accommodation.

188. GARRULUS GLANDARIUS, Linn. The Common Jay.

Spanish. Arrendajo.

The common Jay is very plentiful near Gibraltar in the Cork-wood, and in the wooded valleys and hill-sides up to a considerable elevation. At the same time it is rather local; and though numbers are resident, it seems to be more abundant in the winter months. This bird is not recorded by Favier from Morocco; nor could I obtain any species of Jay on the African side. It would seem, however, probable that they would sometimes cross the Straits, as they occasionally appear at Gibraltar in winter. Four frequented the Alameda and other gardens in the south from about the 10th of November, 1870, to the 4th of April, 1871; and I saw another in March 1872; this last bird did not linger about for more than a few days.

The Jay nests in some numbers in the Cork-wood, laying its eggs early in May; and, at that season particularly, they are easily decoyed within shot by secreting one's self in thick cover and imitating either their call or the squeal of a wounded rabbit.

Order COLUMBÆ.

Family COLUMBIDÆ.

189. COLUMBA PALUMBUS, Linn. The Ring-Dove.

Moorish. Kamoor. *Spanish.* Paloma torcaz.

"This Pigeon is found near Tangier throughout the year. Some are migratory, crossing to Europe in March and April." —*Favier.*

In some localities in Morocco the Wood-Pigeon positively swarms. In April, up a valley near the Foudak, to the south-west of that place, on the road between Tangier and Tetuan, it would have been easy to shoot a hundred in a day, they were in such numbers and so excessively tame. Two or three which we shot to eat, had their crops full of the tuberous root of some weed which had been ploughed up and was lying in quantities about the fallow fields. During the same month, about three years previously, I noticed considerable numbers about Larache; but there they were much more wild, though not so shy as in England or Andalucia. In the latter country, about Gibraltar, a few pairs nest in the Cork-wood and other wooded districts; they are most abundant during the winter months, though I never saw any great quantity.

190. COLUMBA ŒNAS, Linn. The Stock-Dove.

Moorish. Hamam el Berri.

The Stock-Dove is neither mentioned by Favier nor Mr. Drake as occurring in Morocco. I found it near the Foudak at the same time and place that the Wood-pigeons were so abundant. It is sufficiently common to be known to the Moors there by the above-mentioned name, which, by the way, is the same as that used for the next species, *C. livia.* They were in some numbers; and I shot one or two for identification, being further informed by the Moors that they nested in holes of trees. They evidently were breeding at that time; but we failed to discover a nest during the very short period that we remained there. I also noticed the Stock-Dove in April near Larache. On the Spanish side of the Straits, I only observed this species upon one occasion, in the spring, near Gibraltar.

The absence of the white patch on the wing will often serve to distinguish it, when flying, from the Ring-Dove, *C. œnas,* independently of its smaller size, while the absence of the white rump and larger size equally distinguish it from the smaller Rock-Dove, *C. livia*—not that the latter species is often met with in the same locality.

191. COLUMBA LIVIA, Linn. The Rock-Dove.

Moorish. Hamam el Berri. *Spanish.* Zurita.

" This is the most common of the Pigeons about Tangier,

living in rocks and even in the ramparts of the town, breeding both in a wild and in a domestic state all the year round."—*Favier*.

The Rock-Dove is plentiful on both sides of the Straits wherever there are rocks and caves, inland as well as on the coast. Many are resident at Gibraltar, on North Front and at the "back of the Rock;" and at one time some sport was to be obtained upon getting permission to shoot them; but a "young and inexperienced" arrival one day, instead of killing pigeons, shot one or more of the celebrated Gibraltar Apes, for which he, amongst suffering other indignities and punishments, was afterwards known as "Du Chaillu." In consequence of this exploit, all leave to shoot was thenceforth withheld, and very rightly so.

The distinguishing marks of the Rock-Dove are the white rump and two dark bands on each wing. Pied and white varieties are frequently seen; whether escaped tame Pigeons or real wild Rock-Doves, I cannot say.

192. TURTUR AURITUS, Linn.   The Common Turtledove.
Moorish. Imam, Stitsia.   *Spanish*. Tortola.

" Is a summer resident near Tangier, vast numbers arriving to cross the Straits in flocks during April and May, returning in September and October, then to retire south for the winter. This species is without doubt the origin of the domestic Turtle-dove, called *limama* or *dekrallah* (praise of God), some of which birds are pure white."—*Favier*.

The common Turtledove is seen in extreme abundance in Andalucia, during its stay being a great object of pursuit to the Spanish *tirador*, who, in August, often makes a *puesto*, or hiding-place, near some favourite drinking-haunt of the doves, and shoots them much in the same way as the Partridge—that is, on the ground, three or four in a row; only with these birds he has, of course, no *reclamo*, or call-bird.

The Turtledove chiefly migrates during the first week in May, more arriving in that week than during all the rest of the time of their migration. I first saw one in 1870 on the 11th, and in 1872 on the 14th of April. They mostly disappear by the beginning of October. The latest I ever noticed was a single bird at Casa Vieja on the 31st of October. In my note-book I have one down as seen on the 9th of October, in the middle of the Bay of Biscay.

193. TURTUR SENEGALENSIS, Linn. The Egyptian Turtle-dove.

This species is mentioned by Mr. Drake as found commonly in the southern part of Morocco, but does not appear to have been obtained by Favier in the north-west of that country.

The absence of spots on the back, and its smaller size, will distinguish it from any other species likely to be met with in Morocco.

Family PTEROCLIDÆ.

194. PTEROCLES ARENARIUS, Pall. The Black-bellied Sand-Grouse.

Moorish. El Koudri. *Spanish.* Corteza.

Favier merely mentions that this Sand-Grouse "occurs in Dar el Baidar."

On the European side, this large species of Sand-Grouse, though extremely local, is resident in the marismas and near Utrera, nesting late in May, but does not appear near Gibraltar. I do not think there is any migration of this bird. The different species of Sand-Grouse only lay three eggs, of an *elliptical* form, placed on the bare ground without any nest; eggs of this species taken near Seville are of a pale cream-colour, marked all over with faint spots of very light brown.

195. PTEROCLES ALCHATA, Linn. The Pin-tailed Sand-Grouse.

Moorish. El Găta. *Spanish.* Ganga.

Favier states this species to be "scarce near Tangier, but common about Dar el Baidar;" and he says "they cross the Straits in spring, returning in August and July."

This beautifully marked bird is the most common Sand-Grouse on the Spanish side, although very local, being abundant about the edges of the marismas, where they nest late in May, as also near Granada. Some may be migratory; but I have seen others in January near Seville. I never saw any in the neighbourhood of Gibraltar.

The flight of both species of Sand-Grouse is very powerful; and sometimes they go to such a height that, although you can hear their croaking hoarse call, they are almost out of sight. Becoming excessively tame and familiar when kept in confinement, in a wild state they are very difficult to approach without a stalking-horse, and when obtained are of no use to eat. They are very difficult birds to skin, the feathers coming out like those of a Pigeon.

Eggs of this Pintailed Sand-Grouse, taken near Seville, are of a reddish buff colour, marked all over with spots of reddish brown and light grey.

Future visitants to Morocco should look out for *Pterocles senegallus*, Linn., which most probably occurs in the southern part of the country.

Order GALLINÆ.

Family PERDICIDÆ.

196. CACCABIS PETROSA, Gm. The Barbary Partridge.

Moorish. Hedjel.

"The Barbary Partridge," says Favier, "is resident around Tangier, and very common, sometimes perching on trees."

This species is far more common in proportion in Morocco than *C. rubra* is in Spain, and chiefly frequents palmetto scrub; in some localities it is so numerous that it would be quite easy for one gun to bag fifty brace in the day. The flesh of this Partridge is not so good even as that of the common Red-leg, which does not say much in its favour. They submit to captivity very well, and may be kept alive in coops like fowls, to be used as required—and after being fed on corn for a month or so, improve greatly in a culinary point of view.

As is well known to all ornithologists, the Barbary Partridge is the only species found on the Rock of Gibraltar, being in great numbers there. Sometimes they may be seen sitting on the stones within a few feet of the sea. A pair or two used always to frequent the rocks below the "rope ladder" at Europa. Although protected from guns and carefully preserved from the attacks of human beings, they suffer considerably from the number of cats which abound, and are also preyed on by Genets and Eagles, the Lizards and Snakes destroying the eggs and young.

This bird, like other Partridges, is very noisy at dusk; in the nesting-season they have a peculiar long-drawn croaking cry, which puzzled me for a long time before I could make out from what bird it came. Whether the male only thus calls I do not know; but I suspect such to be the case.

The Barbary Partridge commences to lay about the 15th of April. The eggs are very similar to those of *C. rubra*, and vary much in the markings, some being quite free from the usual small freckles.

The Rev. John White mentions this Partridge as being plentiful at Gibraltar about 1770, and not being found on the mainland of Spain.

This species is at once distinguished by the *chestnut collar* round the neck, studded with small white spots, and is also a smaller bird than the common Red-leg, besides having a metallic blue tinge on the wing-coverts. The legs are not always red, sometimes being a pale buff colour.

197. CACCABIS RUBRA, Linn. The Common Red-legged or French Partridge.

Spanish. Perdiz.

This Partridge is, throughout Andalucia, plentiful and resident, frequenting the *monte* or scrub, not, as in some parts of England, being found in cultivated places. Never known to occur on the African side of the Straits, it is not even found on the Rock of Gibraltar, which would seem rather strange, as it is to be seen on the Queen of Spain's Chair, and occasionally on the plain below within a couple of miles of the neutral ground.

Almost every Spanish sportsman, or *cazador*, keeps one or

more of these birds as call-birds (*reclamos*), each wretched Partridge being confined in a cage which is so small that the unfortunate bird has scarcely room to turn round. To add to this cruelty, at certain seasons they are never given water, as it is supposed to be fatal to them: but in a wild state they drink a great deal; and during the scorching month of August, and the first half of September, one of the favourite Spanish methods of shooting them is to make a hiding-place (*puesto*) near their drinking-haunt, placing call-birds on each side of the water out of the line of fire—so that, when a covey comes to drink, as many as possible may be mowed down at once by the concealed "sportsman," who, throughout the whole year, regardless of the season, shoots them whenever he can, the acme of his diversion being to shoot a Partridge from the nest. However, it may as well be mentioned that these men shoot for profit, not for sport. In spite of this ceaseless persecution Partridges do not decrease, which is truly wonderful.

Light-coloured and white varieties of the Red-legged Partridge are not unfrequent: for some years in succession there was a white covey near the Guadiarro on the road to Gaucin. This species breeds in May. The eggs vary greatly in size, and are subject to the same variations as those of *C. petrosa*; they are usually larger than the eggs of that bird, but are sometimes smaller, thus varying greatly in size.

Mr. Drake mentions a Francolin as occurring in the south of Morocco. I never could obtain a specimen for actual identification, but have no doubt the bird is *Francolinus bicalcaratus* (Linn.). They are to be found near Rabat; and the local name was told me as "*Ragh*."

198. COTURNIX VULGARIS, Fleming. The Common Quail.
Moorish. Summin. *Spanish*. Codorniz.

Favier states that the Quail is very abundant on passage on the Moorish side of the Straits, many remaining to breed, the majority crossing over to Europe during March and April, returning in October and November.

On the Spanish side of the Straits it appears to me that the chief vernal migration of the common Quail is during the months of March and April, whilst the autumnal passage is almost entirely executed during the latter half of September, at that time their numbers being sometimes almost incredible.

The Andalucian cazadores profess to recognize two kinds of Quail—those which are migratory and called "*Criollas*," and those which are resident and so named "*Castellanas*." There is certainly much difference in the colour of the plumage and of the legs, the Criollas being lighter-coloured and slightly smaller birds than the Castellanas, which are very dark; otherwise, in habits, note, and eggs, there is no difference, although at a glance the resident and migratory races can be easily distinguished.

There are a great number of these resident Quails, which, throughout the winter, seem to collect together and haunt certain favourite spots, these places never being without Quail. You may kill three or four and hunt about unable to find more; but go to the same place in a few days' time, and you will find that some fresh ones have taken possession of the ground.

In summer Quail are universally distributed all over the cultivated country; in autumn the best place to shoot them is in the maize-fields or, rather, stubbles. Vast quantities are caught in the spring with small nets by the aid of the " Quail-call" (*pitillo*). The birds begin to call their love-note about the 9th of March; after that time their "*quit que-twit*" is to be heard on every side as long as the nesting-season lasts. They commence to lay in May; and I have known of a nest with eleven eggs taken as late as the 17th October.

Family TURNICIDÆ.

199. TURNIX SYLVATICA, Desfont. The Andalucian Bush-Quail.

Moorish. "Zerquil" (*Favier*). *Spanish.* Torillo.

"This little Quail is both resident and migratory in the vicinity of Tangier, and is a much less common bird than the ordinary Quail, *C. vulgaris*; those which migrate pass northwards during May and June, and are seen on the return passage in September and October. They nest in July, depositing four eggs in any slight depression of the ground, often among corn. The young, from the moment of exclusion, are attended by both male and female—all remaining together in parties for some time, in the same manner as Quails. I kept a female bird in captivity for about thirteen months, feeding it on millet and water. This bird was very fond of eating flies, and also used to devour the ants which came into its cage to carry away the dead. Very gentle in its character, the call of this bird was very *triste*: it cooed day and night, much in the manner of a Turtledove; only the note was more subdued and lengthened. I have reason to believe that these Quails would breed in captivity, although this individual bird did not lay."—Favier.

On the Spanish side I was unable to detect any migration of this Bush-Quail, though it is said by Andalucian bird-catchers and cazadores to be migratory. The probability is that they are so; but yet I am inclined to think the reverse, as they are found in the same localities in equal numbers at all seasons of the year.

Near Gibraltar it is a very local bird and nowhere plentiful, apparently less so than is really the case; for they are difficult birds to flush, and if put up once will rarely rise a second time. Scattered here and there, they chiefly frequent palmetto scrub and appear to be most common near the coast, being more abundant to the east of the Queen of Spain's Chair, especially about the Lomo del Rey and a place called Los Agusaderas. In their flight and habits, from what I could observe of them, they resemble the Indian Bush-Quail (*Turnix dussumieri*).

I have often seen them among the rough grass and bents close to the sea-shore. One bird in particular, I remember, for a long time frequented a patch of thick herbage near the mouth of the "First River;" and whenever I rode by, my dog used to flush it, till at last one day, wanting a specimen, I went purposely to shoot the bird; but, of course, upon this occasion my friend was not to be found, nor did I again see one there for some months.

They are scarce between Algeciraz and Tarifa, but occur

towards Vejer, and are tolerably plentiful on the palmetto-covered high ground above Casa Vieja, called La Mesa; further than this I did not meet with it personally; nor could I obtain any near Seville.

The nest is, from the skulking habits of the birds, extremely difficult to obtain. I never had the good fortune to find one, but had one lot of eggs brought to me from near San Roque on the 6th of July, 1869. The finder said the nest was under shelter of a palmetto bush, and merely consisted of a few bits of dried grass. These eggs, four in number (which is, without doubt, the regular complement laid by all the Bush-Quails, *Turnix*), were very slightly incubated, and in appearance much resemble those of the common Pratincole, *Glareola torquata*, only being, of course, much more diminutive. Later in July I received several eggs from Mogador, which exactly resembled the Spanish ones; but not having been blown and being hard sat-on, the shells were so tender and rotten that I could do nothing with them. My friend Mr. Reid, of the Royal Engineers, informs me that he had the luck to find a nest, placed in grass near the shore on the eastern beach, on the 19th of May, 1873; this nest contained four eggs (incubated), as did another from near Tangier obtained by Olcese.

The males of this species, and, I believe, of all the genus, are very much smaller than the females; this difference is so striking that the cazadores always declare there are two species. I have at different times kept these little birds alive, and sent one to England, and at present have a male alive. They are easily reconciled to captivity, and become very tame and confiding pets; at times they coo in a moaning way, whence their trivial Spanish name of *torillo* or little bull. They also have another single note, much like that of the female Quail but less loud.

This species has no hind toe; so it has been called by the unnatural and pedantic name of Hemipode. That of Bush-Quail, or, as used in India for other species of the genus, Button-Quail, or even Three-toed Quail would be much more appropriate.

Order GRALLÆ.

Family Rallidæ.

200. Ortygometra crex (Linn.). The Landrail, or Corn-Crake.

Moorish. Zelga (*Favier*).   *Spanish.* Rey de los Codornices (King of the Quails), Guia de los Codornices (Guide of the Quails).

"This Crake is found in Morocco on passage, crossing the Straits during the month of February, returning in August, September, and October, being occasionally obtained throughout the winter months."—*Favier.*

The Landrail does not seem to remain in Andalucia during the breeding-season, as I never heard its well-known cry; but I have seen it as late as the 2nd of May. It is not obtained in any abundance, but, like other Crakes, is, no doubt, more common than it appears to be. It occurs most frequently in October and February, and, as Favier states concerning it in Morocco, is found during the winter.

201. Porzana minuta, Pall. The Little Crake.

The Little Crake is not noticed by Favier as occurring in Morocco; and on the Spanish side of the Straits I never could succeed in meeting with it, though I have seen specimens said to be Andalucian. Owing to the powers of concealment which these small Crakes possess, it is very difficult to obtain them, and impossible to learn much of their habits. Without the aid of a good dog it is very hard to compel them to rise; and in consequence they appear to be much more rare than they really are. I recollect finding a nest of this or Baillon's Crake situated in a very small isolated patch of swamp; and instead of trying to snare the birds, I stupidly endeavoured to flush them with three good water-dogs: but it was quite useless; we could find no signs of them whatever; so the unidentified eggs were valueless, as the resemblance is so great between the eggs of this and

Baillon's Crake, that unless the bird be obtained it is impossible to tell to which species they belong.

The Little Crake has no white marks on the wing-coverts, and very few on the centre of the back, and is slightly larger than Baillon's Crake.

Length 7 inches.

202. PORZANA MARUETTA, Leach.   The Spotted Crake.

Spanish. Polluela (under which name the next four species are also included).

"This bird is met with near Tangier during passage, but not in any great number, and is the most common of the family, haunting thick beds of rushes in swamps and on the edges of lakes and rivers."—*Favier.*

The Spotted Crake is extremely abundant on the Spanish side, being more numerous than the Water-Rail, and is most frequent in spring and autumn. Many remain during the winter months; and they occur also sometimes in the breeding-season; so, although I did not actually obtain an identified nest, I have no doubt they are to be found breeding in the country.

203. PORZANA PYGMÆA, Naum.   Baillon's Crake.

Favier says of this bird:—" Very rare; I only met with one, in 1857."

Seldom obtained, owing to its skulking propensities. I found this prettily marked Crake very common when snipe-shooting at Casa Vieja from October to February. We also obtained it at the Laguna de Janda in May. Many are resident, breeding at the end of April, when they make a small nest of sedges and grass placed at the edges of swamps, laying from five to seven olive-brown eggs spotted with darker brown. It is a smaller bird than the Little Crake, and further distinguished from that bird by the numerous white marks on the centre of the back, scapulars, wing-coverts, and inner secondaries.

Length $6\frac{1}{2}$ inches.

204. RALLUS AQUATICUS, Linn. The Water-Rail.

"This bird is found on passage near Tangier in about the same numbers as the Landrail, frequenting the edges of rivers and swamps, where they hide up in the sedges."—*Favier*.

The Water-Rail is very common in all suitable localities on the Spanish side; and their croaking frog-like call is always to be heard in the swampy jungle at Casa Vieja. Being to a great extent a migratory bird, it is most common in winter; but, owing to the cover being more thin, at that season all the Rails and Crakes are easier to obtain. They build in rushes or sedges, laying about the 20th of April. On the 13th of May we found two nests, from each of which Mr. Stark succeeded in snaring one of the old birds; these nests, built entirely of dry sedge and lined with a few bits of dry grass, were just raised above the water, and measured 6 inches in height, depth, and diameter; the hollow of the nest was $4\frac{1}{2}$ inches across by $2\frac{1}{2}$ deep. Each nest contained seven eggs hard sat-on—one lot being of the usual type, the other resembling more those of the Spotted Crake, or, rather, looking like miniature Waterhen's eggs with larger blotches than usual.

The males are much larger than the females.

205. GALLINULA CHLOROPUS, Linn. The Waterhen or Moorhen.

Moorish. Zelga-kahal (*Favier*).

The Moorhen, according to Favier, is "resident in the vicinity of Tangier, being met with in abundance; many, however, are migratory."

It is needless to say much about a bird so well known as our common English Waterhen. It is not so common in Andalucia as the Spotted Crake (*Porzana maruetta*); but I was unable to detect any migratory habits on the Spanish side of the Straits, where it is tolerably plentiful and generally distributed in all suitable localities, often being seen about the gardens at the edge of the small stream at Algeciraz and at Vejer, seeming, as in England, to be fond of living in the vicinity of houses and cultivation. They nest about the end of April.

206. FULICA ATRA, Linn. The Common Coot, or Bald Coot.

Moorish. El Ghor *(Favier).* *Spanish.* Mancon.

"This Coot is resident near Tangier, but is not very numerous, often consorting with *Fulica cristata*. Some are migratory, passing northwards during the months of January and February, and returning in August and September."— *Favier.*

I found the common Coot abundant near Tetuan in March; and it is a common bird on the Spanish side, particularly in winter, when very large flocks appear. It nests about the middle of April in all large swamps, particularly at the Laguna de la Janda, where, though the nests are numerous, it is almost impossible to see the birds, owing to the density of the grass and rushes.

207. FULICA CRISTATA, Linn. The Red-lobed or Crested Coot.

"This bird is both resident and migratory near Tangier. Those which migrate return from the north in September. The nest and eggs resemble those of *Fulica atra*, with which species they associate, but are much more numerous."— *Favier.*

This Coot breeds at Ras Doudra in numbers about the 20th of April; and, as above mentioned, the eggs are not to be distinguished from those of the common Coot (*F. atra*); so, unless the bird be snared on the nest, the eggs cannot be said to be identified. I never saw this species in Andalucia, where, however, it is stated to occur, and doubtless is met with further eastward, as I have seen specimens at Granada marked as Spanish. I should prefer to call this species the Red-lobed Coot, as it certainly is not crested, being in all respects similar to the common Coot (*F. atra*) except that it has *two red lobes* on the white frontal patch.

208. PORPHYRIO HYACINTHINUS, Temm. The Purple Waterhen.

Moorish. Kazir (*Favier*).   *Spanish.* Mancon azul, Calamon.

"This bird is chiefly migratory, and is not common near Tangier, passing north during the months of February and March, and returning in December and October. They are occasionally to be seen during the month of January, but not every year. Those which remain for the breeding-season construct their nests in the midst of wet sedges or rushes, depositing (in April) from three to five eggs. When these birds are moulting they are very easy to obtain, as they lose all their quill-feathers at once, and so cannot fly."— *Favier*.

The Purple Waterhen (a large and very handsome bird) is, on the Spanish side of the Straits, very irregular in its appearance both as to time and locality. In some years, during January and February, they are to be seen near Gibraltar in situations where they do not occur at any other time, and are then, doubtless, on migration.

In wet seasons they nest at Casa Vieja in April, in the Soto Malabrigo, in which marsh I have shot them as late as the 27th of October. It is a very difficult bird to flush without a dog; when they rise they make a flapping noise, and with a heavy flight merely take refuge in the nearest thick patch of rushes or wet sedgy jungle, whence, from being Crake-like in their habits, it is almost impossible to make them rise a second time. They are not to be met with except among thick wet rushes. Some are to be found in a few places at the edge of the marismas of the Guadalquivir. The nest resembles that of the common Coot; and the eggs, which are richly coloured, are laid towards the end of April.

The gizzards of those which I have examined contained nothing but vegetable matter (grass, seeds of rushes, &c.), with a good deal of coarse gravel.

Family OTIDIDÆ.

209. OTIS TARDA, Linn. The Great Bustard.

Spanish. Abutarda.

Favier states that this Bustard " occasionally migrates to Morocco during winter from the European side of the Straits, but very rarely remains for the breeding-season."

I have seen one or two specimens obtained near Tangier; and Mr. Drake also mentions another; but it is said to be very scarce. The absence of specimens, however, is hardly a proof of its rarity, as it is in winter such a difficult bird to approach.

On the Spanish side of the Straits the Great Bustard is first to be met with near Gibraltar on the plain below Facinas, about ten or twelve miles from Tarifa; northwards from there it is to be seen in gradually increasing numbers all along the veja of the Laguna de la Janda up to Casa Vieja and along by Medina Sidonia to the plains which lie towards Jerez and the marismas of the Guadalquivir. It is found in abundance along the line of railway to near Utrera, being more plentiful in the open corn-growing country about Marchena, Coronil, and Carmona than in any other district that I have visited; thence to the north side of the Guadalquivir it is also common, particularly about Brenes and Alcala del Rio, sometimes appearing very near to Seville: in fact in all open country the Great Bustard may be expected to be seen in varying numbers. I never saw any very large flocks; occasionally I have observed as many as fifty together; but from ten to twenty-five is the usual number seen at once. I never could detect any migration of the Great Bustard, which is singular, as in the Crimea I recollect very large flights appeared on passage in autumn. Perhaps it is that in Andalucia they are always able to obtain sufficient food. In a wild state they feed chiefly on grass and vegetable substances; and when kept tame they will devour any amount of grasshoppers, insects, &c., the best food for them, however, being cabbage-leaves.

Bustards are very difficult to approach except by some such stratagem as driving a cart near them, when they seem to fear no danger; but the best way of obtaining them is by driving when the corn is sufficiently high to shelter the guns, which it usually is by the end of March. It is necessary to have for a Bustard-drive, with any chance of success, at least four guns; the more the better; and as the birds fly almost always well within shot of the ground, they are very easily killed if they pass over the spot where a gun is posted. Indeed, considering the size of the bird, it is wonderful how light a wound will bring one down. I have seen an old male when winged and, as it were, brought to bay, turn round and charge his pursuer. This diversion of Bustard-driving is rather expensive sport, and often, like the Irishman's pig, they refuse to be driven in the required direction; so, beyond the novelty of the affair, and the sight of so many of these truly noble birds on the wing, there is nothing very exciting in the sport; and, as in all "driving," there is no sporting-skill required on the part of the shooter. Nevertheless the thorough enjoyment of the bright and glorious climate, and the sense of freedom to go where you wish without being warned off as a trespasser, and, last, but not least, the sociable nature of the "entertainment" render a few days' Bustard-driving very agreeable. There is, however, the lamentable fact that the game is not very much worth having, the flesh being dry and coarse; at least such has been the case with almost all that I have tasted.

One circumstance in favour of these excursions after Bustard is that they are easily managed from Seville by starting by the early morning train and returning late in the evening, and there is no trouble in having to search for uncomfortable country quarters for the night.

The man we always employed to drive was one Molino, of Algaba, a small village or pueblo on the Guadalquivir, above Seville, a wiry active little fellow, but with an enormous capacity for meat and wine. He never attempted to drive the birds with more than three men, including himself; but his skilful management, owing to his knowledge of the birds and the ground, and consequently of the route they would take, was something marvellous. Molino is always employed by some Sevillanos, who regularly several times in the spring go out dressed in green, like Free Foresters at the Crystal Palace: and a suitable dress it is; for the colour being that of the corn or grass in which the *tirador* lies hid, the Bustards are not nearly so likely to notice the ambuscade. The day these verdant gentlemen choose for their "funcion" of Bustards is invariably Sunday; and sometimes they succeed in killing a dozen birds in the day, usually about the vicinity of Las Alcantarillas. In August, near Casa Vieja, and, I am informed, also in other places, the Spaniards ride down Bustards with dogs, continually flushing them till they are exhausted; but probably young birds only are thus caught. They are also said to tire out the Red-legged Partridge in the same way. This is very likely, as I have seen these Partridges in Norfolk, after being flushed two or three times, allow themselves to be caught when quite uninjured.

The Great Bustard is easily noticed when on the ground where the cover is not too high to hide them; and at times their size appears gigantic as they fly with a slow, measured, laborious-looking flight; but their pace is much faster than it appears to be; and when put up they often fly a distance of at least two miles. They have great power of concealment; and I remember an instance of one which was unable to fly, from some injury he had received in one of his wings. We saw this bird in a corn-field of some forty acres, and forming line we tried to catch him; but he suddenly disappeared in the corn, which was not more than two feet high, and in many places not a foot high. We spent an hour in vain hunting for him with a dog; so, after beating the whole field over more than once, we sat down in view of the ground to eat luncheon. In about a quarter of an hour the Bustard appeared some three hundred yards off in the middle of the corn; so I went straight at him, running as hard as I could. He again disappeared; but going on I suddenly spied something white running, as it were, close to the ground. I rushed after it, when up jumped the Bustard, running along and flapping his wings. As I could not catch him (for he ran as fast as I could), I was compelled to shoot him, a magnificent *Barbon* of about thirty pounds weight. I could not have believed so large a bird could crouch so low and at the same time make such good running.

About Casa Vieja a few Bustards are to be found near the banks of the rivers Barbate and Celemin, where, from the nature of the ground, which is intersected by the windings and branches of these rivers, which are nearly dry in August, it is sometimes possible to stalk them under cover of the banks; and it is very good ground to have them driven over; but the Spaniards there cannot be made to understand such work.

The Great Bustard nests in corn or grass early in May, laying two olive-brown eggs marked with spots and blotches of dark brown. They have been supposed to be polygamous; but I do not think such is the case; the Little Bustard undoubtedly pairs. The gular pouch, which always exists in the old males or *Barbones*, is sometimes very large and necked in the middle, somewhat like an hour-glass, the lower part being the largest; this shape is not constant. Perhaps it is the result of extreme age.

Average length of a male 45 inches, of a female 36.

210. TETRAX CAMPESTRIS, Leach. The Little Bustard.

Moorish. Bou-zerat (Father of the armourer: Râd, "thunderer"), Sáf-sáf. *Andalucian.* Sison, Francolino.

"The Little Bustard is abundant in the vicinity of Tangier in small flocks, which are very wild and wary. They migrate to the north during the months of April and May, returning in October and November. In addition to these migratory birds, great numbers are resident during the nesting-season. The males do not attain the full breeding-plumage until their third year, and by October regain the dress of the females."— *Favier.*

My experience of the breeding-plumage of the Little Bustard is rather different from the above; for, as far as I have been able to ascertain, the males lose the black markings of their nuptial dress by the end of August, if not before. I could not make out the exact period; but I never saw a black-marked male which had been killed after the middle of August.

The adult males never lose the minutely marked or vermiculated plumage on the back, which part in the females and young males is more spotted or blotched, like the feathers of the Great Bustard. I found the Little Bustard equally common in Morocco as in Andalucia on all open low cultivated ground. On the dead level, or vega, of the Barbate near Casa Vieja at times, in early autumn, they positively swarm, in flocks sometimes of as many or more than a hundred together, frequenting this flat ground till it is swamped by the rains. They then resort to drier and higher ground, and these large flocks gradually disperse and break up into lots of from five or six to twenty in number. They are, as Favier remarks, exceedingly wary, except during the breeding-season and in the month of August. At other times the only way to obtain them is by driving, which is very uncertain work, as, unlike the Great Bustard, they usually rise high up at once, and their power and rapidity of flight is astonishing for their size and weight.

They are often to be seen flying somewhat like Golden Plover, twirling and twisting about at a great elevation; and sometimes I have watched them rise and go to such a height that it would have been difficult to tell what birds they were unless I had seen them fly up from the ground.

During August, when it is very hot, between eleven and four, they lie "like stones" in the long grass, requiring a dog to flush them; but the heat is then so excessive that one is almost as likely to get a sunstroke as a Little Bustard, and I myself could never stand such work.

The nearest place to Gibraltar that these birds are seen in is on the plain between Los Barrios and Palmones, where occasionally in autumn and winter a few appear; but they are too much bullied by Gibraltar sportsmen to remain there long. The Moorish names given above are all significant of the rattling noise which the Little Bustard makes in rising; and when the flock is large this can be heard a very long way off. There is none of this sound of the wings in the rising of the slow-flying Great Bustard.

When on the wing, the Little Bustard, except when at a

great height, may always be recognized by its white or, rather, pied appearance, caused by the greater part of the wings being white. When these are closed, and they are settled on the ground, this white does not show, and they are very difficult to make out or notice, particularly as they usually frequent ground which has some cover (in the shape of weeds, thistles, or grass). In the breeding-season they keep entirely among thick herbage, and at that time I never could get a sight of one on the ground.

The male Little Bustard in the breeding-season has a most peculiar and offensive-sounding call, which can be easily imitated by pouting out and pressing the lips tight together and then blowing through them; the birds when thus calling seem to be close to you, but are often in reality half a mile off. They must possess powers of ventriloquism, as I have often imagined that they were quite close to me, and upon hunting the spot with a dog found no signs of them anywhere near; indeed, at that season it is sometimes as difficult to make them rise as a Landrail. They nest in the beginning of May, laying three shiny smooth olive-green eggs, more or less blotched with dark brown, which are placed among the corn or long grass. The males in breeding-plumage have the throat and cheeks bluish black, and the breast black, with two white gorget-marks across it.

There is little or no difference in the size of the sexes, the average length being about 17 inches.

211. EUPODOTIS ARABS (Linn.). The North-African Bustard.

This large Bustard was obtained by Mr. Tyrhwitt-Drake in the north of Morocco; and towards the south, about Mogador, it is stated to be common.

Larger than the Great Bustard (*Otis tarda*), it has the entire back covered with those delicate vermiculated feathers, sandy brown crossed with fine lines of deep brown almost black, which are so valuable for artificial flies. Similarly marked feathers are found in the Indian and Cape species (*E. Edwardsii* and *E. cristata*), as well as in several others of the Bustard family.

212. HOUBARA UNDULATA (Jacq.). The Houbara or Ruffed Bustard.

This Bustard is not mentioned by Favier; but I saw one specimen of it, which had been obtained near Tangier in August; further south it is stated to be frequently met with.

Though the Houbara has doubtless occurred in Andalucia, it must be considered extremely rare there. I never saw or heard of one.

It is easily to be distinguished by the white crest and by the black-and-white ruff formed by the elongated feathers which grow at the back of the neck. Whether this plumage is peculiar to adult males I cannot find out; but probably such is the case.

Length about 24 inches; tarsus 4 inches.

Family ŒDICNEMIDÆ.

213. ŒDICNEMUS CREPITANS, Temm. The Thick-knee, or Stone-Curlew.

Moorish. El Karuana. *Spanish.* Alcaravan.

The Thick-knee is found on both sides of the Straits as a resident in considerable numbers, nesting generally about the beginning of May, and depositing its complement of two eggs usually on stony dry ground.

These birds are far more common in the winter months, and most so during their migration, which is northwards during March and April, and southwards in October, November, and December. They pass in lots of from five or six to fifty in number, and are chiefly observed on ploughed fields, generally near the banks of rivers, where I have sometimes shot them as they flew by when I was waiting for ducks in the evening. Except in the breeding-season, I have found them rather wild. They are, when nesting, very noisy at night, and are, doubtless, nocturnal-feeding birds.

Family GLAREOLIDÆ.

214. GLAREOLA PRATINCOLA (L.). The Collared Pratincole.

Moorish. Gharrak (*Favier*).   *Spanish.* Canastela.

Favier's notes of this Pratincole are confined to remarking "that it arrives from the south and passes to Europe during the month of April, being observed returning thence in September to join those which have remained near Tangier for the breeding-season. All disappear south for the winter months."

I found this bird in April, on the dried mud at the lakes of Masharal hadar, south of Larache, in countless thousands. They had not then begun to lay; so possibly some of these swarms would pass on northwards. I there witnessed a number of these birds mobbing a Marsh-Harrier which had intruded on their ground, buffeting and bullying him just as Peewits will do when a Hawk passes near their breeding-ground. At times at least a hundred Pratincoles were dashing at once about the Harrier, which soon made its best way out of their district. Pratincoles are very crepuscular in their habits, flitting up and down over the surface of a river or a pool much after the manner of a Skimmer (*Rhynchops*). They fly very late in the evening—as late, indeed, as they can be distinguished. They are then silent; but by day, especially when disturbed, their cry is ceaseless; and the Moorish name given by Favier is doubtless derived from, as it is suggestive of, their note. They are generally very tame and fearless, often allowing one to approach within a few yards. They are birds of very powerful flight, and remind one much of the Terns in this respect.

On the south side of the Straits the Pratincole is found in large numbers wherever there are lagoons, which, drying up in spring, leave a surface of sun-baked mud on which they deposit their complement of three eggs only about the second week in May. The earliest egg I saw was taken on the 3rd of that month. They mostly arrive about the 20th of April, the earliest date on which I saw one being the 4th and 10th of that month in two consecutive years. They fly very high when on passage, and attract notice chiefly from their cry. The latest date of the return migration observed was the 14th of October, when a young bird was procured.

I failed to discover any nesting about the vega of the Laguna de la Janda, the marismas of the Guadalquivir being their chief resort.

A friend of mine, who shot several on the autumnal passage, informed me that they were excellent eating; but in this respect I can give no personal information.

The sexes are alike in plumage; and the species is distinguished by its rufous axillaries.

Family CURSORIIDÆ.

215. CURSORIUS GALLICUS (Gm.). The Cream-coloured Courser.

Moorish. Gueta (*Favier*); but this name applies also to *Pterocles alchata.*

"This Courser appears annually during July in some numbers on the plains of Sharf el Akab, not far from Tangier. Their stay there and their numbers vary according to the abundance or scarcity of insects, and also with the temperature; for unless the latter is favourable, they are rarely met with, and none were seen during the year 1854. They leave these plains in August or the first part of September. Early in summer they ought to be found nesting near Sharf el Akab, as in May 1847 a male was brought to me by a chasseur, who rescued it from a Falcon which had struck it down.

"Their food is entirely insects or larvæ, particularly *Pentatoma torquata* and different sorts of grasshoppers. They are met with in small lots, usually frequenting dry arid plains, where they spread out in all directions, running about after insects, and are very wary and difficult to get a shot at. Their cry of alarm is much like that of the Plover. They rest and sleep in a sitting position, with their legs doubled up under them. Should they not fly away when approached, they run off with astonishing swiftness, manœuvring to get out of sight behind stones or clods of earth; then, kneeling down and stretching the body and head flat on the ground, they endeavour to make themselves invisible,—though all the time their eyes are fixed on the object which disturbs them, and they keep on the alert ready to rush off again if one continues to approach them.

"The age of the young birds can be well made out by the zigzag markings with which the plumage is speckled, which becomes clearer each moult till the end of the second year, when they assume the regular adult livery. There is no difference at any age in the plumage of the sexes.

"In 1849 they did not leave till the 11th of September, when a chasseur brought me one slightly wounded in the wing. I tried to keep this bird alive; but it died directly the weather became cold. It proved on dissection to be a female; and from the large size of the eggs in the ovary it appeared as if it would soon have nested, probably in October or November, when doubtless they retire to a much warmer climate.

"Towards the end of August 1851 two others were brought to me, both slightly wounded—one an adult, the other an immature bird. To prevent the birds this time from dying of cold, I placed them by day in a room where there was always a fire kept up. At night I put them in a box, making a door at the side, lining the top and sides with cotton-wool, placing sand an inch deep on the bottom; this was warmed and dried by putting a charcoal brazier inside during the day. I fed the birds on grasshoppers till November, when these insects became very scarce, and, as each bird ate fifty daily, it was necessary to change their diet to the larvæ of coleoptera, which, after some reluctance, they began to take. This food suited them better than grasshoppers, the birds becoming fatter, at the same time eating less. They did well till January, when, the adult bird pining and refusing food, I tried to save it by cramming; but this was useless, as it died in February, and on dissection I found that death was caused by a very large tumour in the stomach. It proved to be a female; and from the ovaries it appeared the season for laying had passed.

"The surviving bird continued well till the end of January; then, appearing ill, I fed it by hand till April, when as the weather became warmer it grew more healthy. I then shut it up in a cage with a white Turtledove. The Courser was the stronger bird, and did little else than play with the Dove; but they lived in perfect harmony. In May, sexual desire was shown in a very marked manner; but unluckily, the Dove was also a female. During the exhibition of this passion the Courser used to make a noise which may be expressed thus, 'rererer.'

"This continued till the middle of June, then entirely ceasing till the next year (1853), when it resulted in the Courser laying eight eggs—the first on the 15th, the second on the 16th, the third on the 30th of May, the fourth on the 1st, the fifth on the 11th, the sixth on the 14th, the seventh on the 23rd, and the last on the 25th of June. In 1854 she laid again, with the same irregularity, twelve eggs—the first on the 17th of May, the last on the 28th of July. Though in perfect health, treated and fed in the same way, she did not lay in 1855—but in 1856 laid two eggs, on the 6th and 7th of July. In 1857 she again, at irregular intervals, laid ten more eggs—the first in May, the last in July. In 1858 none were laid. In 1859 she produced four more eggs—the first two on the 6th and 7th of July, the others on the 9th and 10th of August.

"Shortly afterwards this bird, in perfect health, plumage, and vigour, was lost to ornithology, owing to the war between Spain and Morocco; for on the 25th of October I was ordered, with other French subjects, to embark in the French war-steamer 'Mouette,' and not knowing when I should return, and still less how to take care of my bird, I made up my mind to let it go; but it was so tame that it either would not or could not use its wings: so, in my dilemma, I gave it in charge of a Moor during my absence; but, unfortunately, on my return in April 1860 I found it had died.

"From my observations it seems that these birds could be

domesticated and bred so as to be perhaps used for the table; but their value would make them rather expensive luxuries, worthy of comparison with those splendid feasts given by the ancient Roman Emperors; for each bird would cost more than twelve dozen capons.

"Meanwhile, considering the eggs laid by the above-named female bird, the size of the testes of the males and ovaries of the females in August and September, one may conclude that they breed more than once a year, and that the complement of eggs is two. These are rather elliptical in shape, of a cream-colour, spotted or marbled with red, bluish ash, and brown."—*Favier.*

Favier is clearly in error in supposing that the complement of eggs laid by this Courser in a wild state is two, as it is now well known that birds of this tribe, like the Pratincoles, usually lay three eggs. Those of the present species are figured in 'The Ibis' for 1859 (p. 76, pl. ii.).

I never had the good fortune to meet with the Cream-coloured Courser on the Spanish side of the Straits; nor is it included by Mr. Saunders in his 'List of the Birds of Southern Spain.' Probably it does occasionally occur; and from what I now know, the most likely locality to look for it near Gibraltar would be the flat ground near Tapatanilla, on the road between Tarifa and Vejer.

Family CHARADRIIDÆ.

216. VANELLUS CRISTATUS, Meyer. The Peewit, or Lapwing.

Moorish. Bibet, Dihoudi. *Spanish.* Ave fria (cold bird).

"This Plover occurs near Tangier in abundant flocks throughout the winter months, arriving from the north during October and November, crossing back again to Europe in February and March.

"The superstitious Arabs believe that these birds are Jews changed into the shape of birds, and also believe that they still retain all their Israelitish characteristics, even wearing the black Hebrew skull-cap."—*Favier.*

I observed towards the end of April three or four pairs of Peewits, which were nesting at the northern end of the lakes of Ras-Dowra. As we had not sufficient time to go further than the commencement of these lakes, probably many others were to be found breeding still further south. The place where I saw them was some eighty miles at the least to the south of Tangier. On the Spanish side of the Straits very few, compared with their numbers in winter, remain to breed in the marismas of the Guadalquivir, where I found the nest with young on the 26th of April. Curiously, none remain to breed about the Laguna de la Janda, or, as far as I could ascertain, anywhere but in the marisma. The majority of the Peewits arrive near Gibraltar about the middle of October, and take their departure north about the first week in March. During the winter they are to be found on every level piece of ground; and I have seen them occasionally on the green glacis of the batteries near the Alameda at Gibraltar, and often noticed them on the "North Front;" while at times they are found on the hill-sides at a considerable elevation if there be any grassy and suitable open spot. They seem to be more scattered and dispersed about than is usual in England, although large flocks may sometimes be seen.

The Peewit used to be an unfailing source of diversion to the British subaltern, and also to the "sportsman" of the Rock, who, a marvel of leather straps, gaiters, bags—leather all over—used to sally out of Gibraltar for a Sunday's shooting, accompanied generally by a bob-tailed, mangy, lean and hungry-looking species of pointer—by its appearance warranted to devour immediately any thing its master might kill; but the unfortunate animal probably seldom had the opportunity of having its appetite so gratified.

217. SQUATAROLA HELVETICA (Linn.). The Grey Plover.

Spanish. Redolin.

Favier only remarks that "this Plover is found near Tangier between the months of December and March."

The Grey Plover appears chiefly to arrive near Gibraltar during the middle of November, and, though frequently seen in autumn and spring, cannot be said to be at any time abundant. I have noticed them on the small plash of water which after heavy rains is formed on the western part of the Neutral ground.

On the 22nd of May, 1869, I killed a pair at one shot near the mouth of the Guadiarro, the male being in almost perfect summer plumage, the female not being so far advanced, and the eggs in her ovaries very slightly developed. It is very remarkable that this northern-breeding bird should linger so late in such a sunny southern country; and I remember the day above mentioned was very hot for the time of the year. But the Grey Plover is not singular in thus remaining south so late; for both the Knot (*Tringa canutus*) and the Curlew-Sandpiper (*T. subarquata*) are equally loiterers late into the spring.

The marks by which this species is to be distinguished from the Golden Plover (*Charadrius pluvialis*) are the presence of a very small hind toe and the black axillaries; it is, moreover, a larger bird, and shore-frequenting, not being found inland like the Golden Plover.

218. CHARADRIUS PLUVIALIS, Linn. The Golden Plover.

Moorish. Tullit. *Spanish.* Chorlito.

"This Plover is very abundant around Tangier in large flocks, which arrive during October and November, and which return to Europe in February and March."—*Favier*.

The Golden Plover occurs as above on the Andalucian side of the Straits; but at the same time their numbers fluctuate very much, in some winters the quantity seen being very great. Upon their first arrival they are generally tame; but being so much sought after by the *cazadores*, they soon learn their danger and become more wary. The earliest I ever noticed them near Gibraltar was on the 1st of November; the latest was on the 6th of March.

The best ground for Golden Plover is the vicinity of Tapatanilla. They always frequent the same places; and if put up from any spot, they are almost certain to return within an hour or two. I regret to say that occasionally the

Spaniards catch them like eels, by laying night-lines or hooks baited with a worm in their feeding-places: this is a most cruel method of procuring them, as the unfortunate bird lingers in agony for hours, often being left till it flutters itself to death.

In Malta the Eastern Golden Plover (*C. fulvus*) has appeared, and may occasionally wander into Spain. The distinguishing mark of the Golden Plover from that species is its *white* axillaries, which are smoky grey in *C. fulvus*.

219. EUDROMIAS MORINELLUS (Linn.). The Dotterel.

"This bird, which appears to travel in company with *Cursorius isabellinus*, is found near Tangier sparingly on its annual passage during August and September, frequenting in small flocks the same dry places that the Courser inhabits; and, like them, it seems to dread the cold."—*Favier*.

Could Favier occasionally have seen the Dotterel on mountain-tops in Scotland, he would not have supposed them to fear the cold; but curiously enough he omits to mention the date of their appearance in spring; and I have no record in my notes of observing it at that season on the Spanish side of the Straits: the few I have seen were in autumn, the latest being shot about the 9th of November. Probably they pass straight on, and thus appear rarer than is the case; but I imagine their line of migration must be further to the east.

220. ÆGIALITIS HIATICULA (Linn.). The Ringed Plover or Ring-Dotterel.

Spanish. Frailecillo, Correrios: these names belong also to the next two species.

"This Ringed Plover is, near Tangier, found in small numbers in pairs and companies on the sea-shore. They arrive during the months of September, October, and November, returning north again in April and May."—*Favier*.

Though I have no absolute proof, I am nearly sure that this species occasionally remains to breed near Gibraltar, as I have shot them as late as the 28th of May, and have seen eggs obtained near Seville as early as on the 23rd of March; but this is the only instance I know of their nesting so far south. During autumn and until April the Ring-Plover is extremely plentiful along the coast, and most so in the month of March.

221. ÆGIALITIS FLUVIATILIS (Bechst.). The Little Ringed Plover.

This small species is not mentioned by Favier in his MS., although Mr. Drake includes it in his 'List of the Birds of Morocco." I found it very common in that country on river-banks during the month of April. Equally abundant at that time on the Spanish side, the Little Ringed Plover is not so frequent in winter as during the breeding-season. They mostly arrive about the 14th of March, some passing on; others remain to nest, depositing, about the 14th of May, four eggs on the sand or shingle by the sides of rivers. Many pairs nest on the river Barbate, near Casa Vieja. There is, of course, no approach to a nest; but the eggs, with the small ends inwards, are placed in a small depression probably formed by the birds themselves in the sand or gravel. They also (like their larger brethren, *Ægialitis hiaticula*) often nest on open flat ground far away from water.

In spring the eyelids are naked and golden yellow, and the legs are of a pale ochre-brown colour; and although its smaller size is sufficient to identify the species, it may be always distinguished from *Æ. hiaticula* by its having the shaft of the *first primary only* white, whereas in the Ringed Plover all the shafts are white.

222. ÆGIALITIS CANTIANA (Lath.). The Kentish Plover.

Moorish. Bou-hejaira (father of stones, *Favier*). *Spanish.* Charran.

"This bird is very abundant near Tangier, and generally found at the mouths of rivers. Many are resident, those which are migratory arriving during September and October, leaving northwards in March and April."—*Favier.*

The Kentish Plover is by far the most plentiful of the seashore Waders on the Spanish side; and they are always very tame, being seldom molested by the Spaniards. The local name of *Charran*, which I have heard for them near Gibraltar, signifies a low unmitigated blackguard; and should this epithet be applied to a Spaniard, the result would be probably a "*puñalada*," or stab with the queer-shaped clasp-knife (*navaja*) which every Spaniard carries.

The Kentish Plover is found throughout the year, but is most abundant during the seasons of migration. They are very active, nimble little birds, running along the shore sometimes in front and within a yard of one's horse's feet; and they are frequently seen running out on the wet sand as one wave recedes, to rush back again as another returns: but they are much too agile ever to be caught by the influx.

About the 20th of April they commence to lay their four stone-coloured eggs, marked with black spots and streaks. Some regularly breed on the dry sandy hillocks and banks near the mouth of the First River (Guadarranque); but, like the other species of *Ægialitis*, they frequently nest far away from the shore, as on the dried mud of the marisma.

The adult males have, in the breeding-plumage, the top of the head and occiput reddish brown, the forehead white bordered by a black patch; the legs, toes, claws, and bill are black.

223. STREPSILAS INTERPRES (Linn.). The Turnstone.

Moorish. Shorno (*Favier*).

"This bird is not numerous, being found near Tangier on the coast and sometimes on the edges of freshwater lakes. They are more abundant near Rabat. Arriving from the north in September, they return in February."—*Favier*.

The Turnstone is found on the Spanish coast in autumn and spring. I never saw them in any quantity, and chiefly observed them about the end of March, occasionally in company with the Ring-Plover (*Ægialitis hiaticula*).

224. HÆMATOPUS OSTRALEGUS (Linn.). The Oystercatcher.

Moorish. Aisha el behar (*Favier*).

"This species is found near Tangier on passage, passing

north during April and May, and returning in October."—*Favier*.

Favier also states that the Oystercatchers sometimes remain to nest. This is possibly the case; but the eggs which were marked as "Oystercatcher's" by him were to all appearance those of the Stone-Curlew (*Œdicnemus crepitans*).

On the Spanish side the bird is not at all numerous near Gibraltar, and appears irregularly from autumn to spring, the latest recorded by me having been one observed on the 5th of May by Lord Lilford near the mouth of the Guadalquivir.

225. RECURVIROSTRA AVOCETTA, Linn. The Avocet.

Moorish. Bou mehet (*Favier*). *Spanish*. Avoceta.

"This species is not common in the vicinity of Tangier, being only met with on passage, on the edges of rivers and lakes, in small flights, which pass northwards during March, April, and May and return south in November."—*Favier*.

I never personally met with the Avocet on either side of the Straits. A few pairs nest in some parts of the marismas during the month of May, and specimens of both eggs and birds are occasionally brought into Seville; but it cannot be, unless very local, a common bird.

226. HIMANTOPUS CANDIDUS, Bonn. The Black-winged Stilt.

Moorish. Bou-ksaiba (*Favier*). *Spanish*. Cigueñela.

"This bird is not found close to Tangier, but frequents freshwater lakes further south, where many remain for the breeding-season; others, arriving during the month of April, pass on northwards and return in November. They appear to migrate by night."—*Favier*.

This Stilt is, in spring, one of the most common of the marsh-birds on both sides of the Straits. At Masharalhaddar, in Morocco, and in the marismas of the Guadalquivir their numbers are perfectly marvellous. I could not find them nesting nearer to Gibraltar than the above-mentioned marisma. They frequent open shallow pools and lakes, and are very seldom seen where there is grass or rushes. They are generally very tame and confiding; while their conspicuous black-and-white plumage and noisy habits render them certain to attract attention, either as they fly with their long pink legs stretched out, Heron-like, behind them, or as they wade about, usually up to their knees, in the shallow water, where they seek their food in the shape of aquatic insects, gnats, and flies.

The Black-winged Stilt is almost entirely migratory; but in some years a few undoubtedly remain behind throughout the winter, as I have seen small lots on the 26th and 27th of November in different years, many on the 22nd of December, and others on the 14th of January. The chief number appear towards the end of March and beginning of April; and they are then not unfrequently seen near Gibraltar at the mouths of various rivers, but soon pass on to their breeding-places, where they nest in colonies and deposit their four eggs on the half-dried mud. I have seen eggs as early as the 28th of April; but the majority lay about the 10th of May.

Family PHALAROPIDÆ.

227. PHALAROPUS FULICARIUS, Linn. Grey Phalarope.

Favier mentions only two specimens of this bird as having been obtained by him near Tangier, in December 1858. Mr. Drake also refers to this Phalarope as having been shot near Tangier during the month of January; while, on the Spanish side, I am able to record one, killed on the 29th of November, 1872, at Tapatanilla, on the edge of the Laguna de la Janda.

Family SCOLOPACIDÆ.

228. TOTANUS CANESCENS (Gm.). The Greenshank.

All that Favier has to say of this bird is that it is met with on passage, "returning south during the month of October to winter probably in the interior of Africa;" and as the Greenshank is recorded by Andersson as common in Damara Land, Favier was not much out in his supposition. He does not, however, mention the date of its vernal migration, which takes place in March, April, and May, the birds being most frequently seen during the latter month. The latest recorded by me was the 22nd of May. I have also noticed them in November and January, but not numerous at any time; in all probability their chief line of migration lies further to the east.

The Greenshank is a very noisy bird, and sure to attract the notice of the ornithologist by its loud whistling cry, which, as is well known, consists of three notes.

The total length is about 13·5 to 14·5 inches; the tarsus 2·3 inches.

Future visitors to the neighbourhood of the Straits should look out for *Totanus stagnatilis*, the Marsh-Sandpiper, which is best described as a miniature Greenshank of about 9 inches in length, the tarsus being 2 inches long. This bird, common in the south-eastern part of Europe and in India, probably occasionally wanders to the west of Spain.

229. TOTANUS FUSCUS, Leisler. The Spotted or Dusky Redshank.

Favier's notes relative to this species are as follows:—" Frequents the vicinity of salt marshes near Tangier during the months of September and October;" but the brevity of his remarks on most of the Waders and aquatic birds would lead one to suspect that Favier, like many other Frenchmen and all Spaniards, had a cat-like antipathy to water.

On the Spanish side of the Straits I never shot a specimen of the Dusky Redshank; but it occasionally occurs in spring and autumn.

This bird is to be distinguished from the common Redshank (*T. calidris*) by its longer tarsus and bill; the latter is also more slender at the tip, and the upper mandible is slightly hooked at the point.

The entire length is about 12 inches.

230. TOTANUS CALIDRIS (Linn.). The Common Redshank. *Andalucian*. Archibebe.

Favier states that " this Redshank is very abundant near

Tangier, in small lots, which frequent the edges of rivers and lakes, and mostly pass northwards during March and April, returning to remain for the winter in September and October. Some, however, remain in the country for the breeding-season."

I found the common Redshank in some numbers at the lakes of Ras Dowra towards the end of April; and they were then evidently beginning to nest. They were not in any thing like the quantity which breed in some parts of the marismas of the Guadalquivir, where they are a little later in breeding than the Peewit, which is there the earliest marsh-nesting bird. In Andalucia this Redshank is, though frequently seen in winter, chiefly migratory, passing north in great abundance mostly towards the middle of April, when many are to be seen and heard shrieking out their double note about the old salinas or abandoned salt-pits at Palmones, near Gibraltar; and a great many fall victims, to appear ultimately in the market; but they are quite unfit to eat.

The total length is about 11 inches.

231. TOTANUS GLAREOLA (Linn.). The Wood-Sandpiper.

This bird is not noticed by Favier as occurring in Morocco; but there were plenty to be seen towards the end of April at the lakes of Ras Dowra and other swamps in that country; and near Gibraltar I have observed it frequently on passage from the 9th of March to the beginning of the month of May.

Being, as far as my observations go, entirely a freshwater Sandpiper, this species most resembles in habits the Marsh-Sandpiper (*T. stagnatilis*).

It can be distinguished by the axillaries, which are white, with a few dusky bars; the legs are pale olive-green; and the tarsus is long in proportion to the body, giving it a "Stilt-like" appearance.

232. TOTANUS OCHROPUS (Linn.). The Green Sandpiper.

Spanish. Lavandera.

"This species is not uncommon in winter around Tangier, frequenting the edges of lakes and the banks of rivers alone or in pairs. They depart northwards during February and March, reappearing by August and September."—*Favier.*

The Green Sandpiper, a bird in which I have from my boyish days always taken a special interest, is in Andalucia, as in England, extremely irregular and uncertain in its movements, changing its ground continually. They fluctuate greatly in numbers; days elapse without seeing a single bird, and suddenly several appear; but they are seldom observed in any greater number than two or three together; generally they are solitary in their habits, and without exception frequent shores of freshwater lakes, ponds, and streams. The loud note of this Sandpiper and the white tail-coverts, which show markedly when on the wing, can hardly fail to cause its recognition. The curious fact, only recently discovered, of its nidification in trees in old nests of other birds, has probably led to its nesting in many countries being overlooked; and who can tell that it may not yet be found breeding in Andalucia?

The Green Sandpiper is most common in the winter months near Gibraltar; and the only month in which I have not seen it has been July; but then I had no opportunity of so doing. The species has greenish black legs; the axillaries are black, with narrow white bars; and it almost always has a strong musky odour.

233. TOTANUS HYPOLEUCUS (Linn.). The Common Sandpiper.

Spanish. Andarios, Correrios.

"This is the most common of the Sandpipers around Tangier, passing north during April and May. They are seen returning in August, September, and October."—*Favier.*

The common Sandpiper in Andalucia prefers the banks of running streams and salt or tidal marshes, being seldom noticed in freshwater marshes or about stagnant water. Near Gibraltar, particularly when on passage in spring, they greatly frequent the sea-coast wherever there is much seaweed thrown up by the tide; and I have repeatedly observed them on rocks, like the Purple Sandpiper (*Tringa maritima*).

The present species swarms about the Straits in March and

April, passing in lots of four or five together. I have no record of any in November, but saw one on the 24th of October and one on the 7th of December; they are not abundant in spring until the month of March, their passage being at its height about the 15th of April. Though I never succeeded in finding a nest, I am almost positive they do occasionally remain to breed, as in the end of May, near the mouth of the Guadiarro, I saw a pair which, from their manner, certainly were nesting; but all my efforts to discover the situation were futile.

234. LIMOSA ÆGOCEPHALA (Linn.). The Black-tailed Godwit.
Moorish. Tchibib (*Favier*). *Andalucian.* Abujeta, Sarseruelo.

"This Godwit is found on passage near Tangier in abundant flocks, migrating to the north during the months of February and March; they are observed returning in August and September."—*Favier.*

Favier also further asserts that this species occasionally remains to breed in Morocco—a statement which is extremely improbable, and therefore must be received with great reserve, though it would not perhaps be more surprising than the fact of the Crane (*Grus communis*) and the Peewit (*Vanellus cristatus*) nesting so far south as they have been proved to do.

The Black-tailed Godwit appears in Andalucia during February, in bands of from four or five to two or three hundred in number, frequenting the grassy marshes or rather inundated ground about Casa Vieja and the marismas. They are very restless, and continually on the move, uttering their loud cries. As they are usually rather wary and difficult to get a shot at, the best chance of obtaining any is either to lie up for them, or, in Spanish fashion, to use a stalking-horse. Their numbers vary considerably in different years; and they do not seem to stay long in the same district, as some hundreds may be noticed one day and hardly any on the next. Their passage continues far into the month of March, by which time they are well advanced in their rufous breeding-plumage; and this ruddy appearance shows much when they are on the wing. The earliest assumption of this dress that I observed was on the 24th of February. Immense quantities are brought into Seville market for sale during March; and the latest I saw there was on the 6th of April. In that district their local name is *Sarseruelo*; but about Casa Vieja they are known as *Abujeta*, evidently a word of Moorish derivation.

I have no note of their autumnal migration; but occasionally they are met with in winter, usually solitary birds, as I killed one on the 5th of December at Tapatanilla, and have seen others now and then obtained in January.

The distinguishing marks of the Black-tailed Godwit are the black tail and white axillaries.

235. LIMOSA LAPPONICA (Linn.). The Bar-tailed Godwit.
Moorish. Tchibib (*Favier*).

"This species is, during passage, nearly as numerous in the vicinity of Tangier as *L. ægocephala*. They arrive from the north during September, and, passing on further south, return and cross over to Europe during the months of February, March, and April. The chasseurs of Larache call this Godwit *Boumeraisa* indiscriminately with the other species."—*Favier*.

The Bar-tailed Godwit, chiefly from frequenting salt marshes and estuaries of rivers, is not noticed near Gibraltar in such numbers as the larger Black-tailed Godwit, and is only observed on passage. The latest date I have of seeing them in spring was on the 10th of May; and the earliest date of its appearance recorded in autumn was the 21st of September; but no doubt they pass much sooner than this. I did not see any during the winter months.

This species is remarkable for the great difference in the size of the sexes, the females being considerably the larger birds.

Its distinguishing marks are found in the tail and the axillaries, both of which are barred with black and white.

236. MACHETES PUGNAX (Linn.). The Ruff (male). The Reeve (female).

Moorish. Habib el tchibib (The friend of the Godwit).

"This species is only observed near Tangier when on migration, crossing to Europe during March, returning in July, August, and September. Those which return in the last days of July still exhibit traces of the breeding-plumage."—*Favier*.

The greater number of Ruffs pass northwards through Andalucia in April; but flocks occasionally occur during January, February, and March, some passing as late as the last week in May. The males, or "Ruffs," are then in their inconvenient-looking nuptial plumage; and although I have not heard of their nesting so far south, it is not unlikely that such is the case.

The total length is from $10\frac{1}{2}$ to 12 inches, tarsus $1\frac{3}{4}$; the axillaries are white.

I would here draw attention to an American species of Sandpiper, Bartram's Sandpiper (*Actiturus Bartrami*), which is a great straggler, and possibly may turn up in the south of Spain. It is described as a bird frequenting dry ground, and by a casual observer might be taken for the Reeve or a Ruff in winter plumage; but it can be distinguished by the axillaries, which are barred with black and white. The total length is from 11 to 13 inches, tarsus about 2 inches.

237. TRINGA CANUTUS (Linn.). The Knot.

Favier merely remarks of this species that it "passes near Tangier in June." If such be the case it must be very early in that month. The Knot is somewhat irregular in its appearance about the vicinity of Gibraltar; and I have rarely met with any, and the few seen have only occurred in April and May. Lord Lilford, however, found them in countless numbers about the 10th of May near the edge of the Coto de Doñana. They were at that time in their fine red or summer plumage, and doubtless *en route* to their as yet unknown breeding-grounds in the extreme north.

238. TRINGA NIGRICANS, Mont. The Purple Sandpiper.

This species, which might perhaps be well termed the Rock-Sandpiper, from its habit of chiefly frequenting rocky and stony coasts, is altogether omitted by Favier as a Moorish bird. Though I never shot a specimen myself, it is not of uncommon occurrence in the winter on the coast, being, if I may use the use the expression, a regular sea-side bird.

The Purple Sandpiper is distinguished by the general bluish lead-coloured tinge of the upper parts of the plumage, and by the dirty yellow colour of the legs, which, being rather short in proportion to the size of the bird, give it a squat or stumpy appearance.

239. TRINGA SUBARQUATA, Güld. The Pygmy or Curlew Sandpiper.

All that Favier has in his MS. relative to this bird is that it "passes near Tangier during the month of April, returning south in September."

The Curlew Sandpiper I never obtained on its autumnal passage; but in some years vast numbers passed at Gibraltar towards the end of April, usually in lots of from ten to twenty in number; they were occasionally mixed up with Dunlins (*T. cinclus*), and were chiefly to be seen at the mouths of rivers, particularly about Palmones. When flying they may be easily distinguished by the white rump, which, when they are on the wing, is very conspicuous. They are in good red or breeding-plumage by the 26th of April; that is to say, the male birds are; but the females are slower in assuming this dress, and probably never become as bright as their mates. About Gibraltar this Sandpiper and others bear the trivial name "pitillo." Lord Lilford informs me that he met with the present species at the same place and time as the Knots (*T. canutus*), and in equal numbers. Curiously, during that spring, Curlew Sandpipers were unusually abundant near Gibraltar, but not a single Knot did I obtain or see. There is, however, not much ground suitable for the various species of *Tringa* in the vicinity of the Rock.

240. TRINGA MINUTA, Leisler. The Little Stint.

This small Sandpiper is not mentioned by Favier as occurring on the Moorish coast; but it is found there from autumn to spring, and I fell in with vast flocks at Masharalhaddar on the 26th of April in company with Dunlins (*T. cinclus*) and Ring-Dotterels (*Æ. hiaticula*); they had then attained their full breeding-dress. On the Spanish side, the Little Stint occurs in like manner; but I never saw any large numbers.

This species is to be distinguished from Temminck's Stint by the outer tail-feathers being ashy-brown; and in habits it is more of a sea-coast or maritime bird.

Total length 5·5 to 6·3 inches, tarsus 0·75.

241. TRINGA TEMMINCKII, Leisler. Temminck's Stint.

This Stint is likewise not referred to by Favier or recorded by Mr. Drake as occurring on the African side of the Straits, where, however, it is of course to be found as on the Spanish side, being common there during the winter and found in small parties of from six to a dozen or more in number. They affect the muddy banks of tidal rivers, especially frequenting the salinas, or salt-pits, a sure locality for them being the now abandoned or unused salinas near Pulmones, between Algeciraz and Gibraltar. They are very seldom seen alone, and are usually tame and easy to obtain. I failed to observe any later than the month of March; but no doubt they further prolong their stay in southern parts.

Temminck's Stint is easily distinguished from the Little Stint (*T. minuta*) by its slightly smaller size and by the *pure white* outer tail-feathers.

Total length 4·5 to 5·5 inches, tarsus 0·75.

242. TRINGA CINCLUS (Linn.). The Dunlin.

Favier remarks that the Dunlin "passes to Europe from the Moorish coast during the months of April, May, and June, returning to winter further south in October and November."

This well-known species, however, is to be seen throughout the winter near Gibraltar, sometimes in considerable numbers. Occasionally they wander far up the rivers some distance from the sea, especially in the spring; and these birds appear to belong to the small race which has been named *T. schinzii*, Brehm.

The majority of these Dunlins arrive in flocks about April and May, when they have assumed their full summer dress with black breasts.

243. CALIDRIS ARENARIA (Linn.). The Sanderling.

"This bird is abundant during migration near Tangier in small flocks along the coast, crossing the Straits during March, April, and May; they are found returning south as late as December. I found numbers near Tetuan in February 1848 at the mouth of the river, where they are known to the Moors under the name of Medronan."—*Favier*.

On the African side I saw large flights of Sanderlings early in April between Tetuan and Ceuta. On the Spanish side they are common from autumn to spring along the sea-shore, the latest I noticed being seen during the first week in May. This cosmopolitan species is distinguished by the absence of a hind toe; otherwise it may be classed among the *Tringæ*; in many of its habits it much resembles the Kentish Plover (*Æ. cantiana*).

244. GALLINAGO GALLINULA (Linn.). The Jacksnipe.

Moorish. Saiga (*Favier*).

This bird is stated by Favier to be "nearly as common in the winter months around Tangier as the common Snipe, arriving during November, and departing northwards in February."

On the Spanish side of the Straits the Jacksnipe is generally distributed throughout the winter, and is extremely numerous about some favourite black muddy spots at Casa Vieja, and in the "ojos," or land-springs, at the edges of the marisma; but it is by no means so plentiful as the common Snipe. Towards the end of February, Jacksnipes assemble together very much; and this gathering of them is a sure prelude to the general departure of most of the Snipes for the north. The greatest number of the present species that I ever saw anywhere was in some of the "ojos" westward of Coria del Rio, near Seville; these circular spots, about ten yards in diameter, are very muddy and sparingly covered with short sedge. Many of them held fifteen or a dozen Jacksnipe; and the often-cited but imaginary individual who is said to have found a single Jacksnipe afford him sport for months, until his friend unluckily killed it, would here, indeed, have been in happy hunting-grounds.

I could not ascertain any good local Andalucian name for this bird. I have heard some; but they were so trivial and varied that I forbear to repeat them. The Jacksnipe is said occasionally to arrive in Andalucia towards the end of September; but my two earliest notes of their arrival are the 24th and the 27th of October near Seville. This species should properly stand in a different genus (*Lymnocryptes*, Kaup) from the common Snipe; but for the sake of simplicity I have placed them here together.

245. GALLINAGO MEDIA, Leach. The Common Snipe.

Moorish. Bou monkar (Father of the bill). *Spanish.* Near Gibraltar, Agachadiza; further north, Agachona.

Favier remarks that on the African side "the common Snipe is found very plentiful around Tangier from the month of October until February," which may be said of it likewise as regards the Spanish side of the Straits; and although better sport is to be had with this (in a sporting sense) king of birds on the Moorish side, the amusement is, as has been already stated, greatly reduced by the want of accommodation and utter absence of comfort; not that there is much of the latter in many places on the Andalucian side. At Casa Vieja, Snipe sometimes arrive as early as the beginning of September. I have heard of a straggler during August; but the greater quantity do not put in an appearance till the end of October and the first week in November. They commence their departure in March; and by the first week in April all have disappeared except a stray loiterer, perhaps a wounded bird. I once noticed one as late as the 3rd of May, having observed it for several days previously in the same situation, and would not shoot it, as I wished to see how long it would remain: this bird did not appear to have any thing the matter with it. I never heard the drumming noise of the Snipe in Andalucia—

though at home in England I have occasionally heard them drumming of an evening in the New Forest as early as the 20th of January, the weather then being unusually mild, and the place where I heard them being their regular nesting-ground.

I have often noticed that, in the marshes both in Morocco and Andalucia, the best ground for Snipe was a spot where sedges and rushes had been burnt during the summer; but the consequent absence of cover in these places rendered it useless to try and walk up to the birds, and the only way was to stand or sit perfectly still in the most favourite spot and await their return. I have more than once taken a chair down and sat in it, waiting for their flight overhead, much to the astonishment of the native population, who could not understand such a proceeding.

The tail of the common Snipe consists of fourteen feathers. Varieties or races of this species, varying in size and colour, have been named *delamottii* and *brehmii*; but in what the distinction consists it is difficult to perceive. *Scolopax sabinii* is now generally admitted to be nothing more than a melanism of the present species.

246. GALLINAGO MAJOR (Gm.). The Great Snipe or Solitary Snipe.

Spanish. Agachadiza real.

Favier only mentions a single specimen of this Snipe as having been obtained by him near Tangier, in 1859. It is, however, included in Mr. Drake's list, 'Ibis,' 1869, p. 153, as twice noticed in March.

The Great Snipe is only met with near Gibraltar on passage, "here to-day, gone to-morrow." I saw two and shot one at Casa Vieja on the 24th of October, 1868; one was killed near Gibraltar on the 17th of October, 1871; and I know of another obtained in April. It is there a well-known bird, but, passing north late in April and early in May, and returning again in September and October, is not very liable to be noticed; and I imagine that their chief line of migration lies more to the eastward.

This Snipe is usually very tame, and, lying closely, shows the external white feathers of the tail very much when rising; and it generally alights again within a short distance, never uttering any sound. It is easily distinguishable from the Common Snipe by its larger size, the underparts, breast, and belly being entirely barred; and, further, the tail consists of sixteen feathers.

· Like the Jacksnipe, this species should stand, in my opinion, in a different genus from the Common Snipe.

247. SCOLOPAX RUSTICOLA, Linn. The Woodcock.

Moorish. Himar el hedjel (The donkey of the Partridge). *Andalucian.* Gallineta. *Spanish.* Chocha.

The Woodcock, according to Favier, is "not abundant around Tangier, arriving during November and departing in March."

Uncertain, both in numbers and as to time of arrival near Gibraltar, in some seasons Woodcocks are tolerably plentiful, as in 1873; in others, as in the winter of 1871–72, they are very scarce. Five or six couple in the day for two guns is a very fair bag; but I knew an instance of a Spanish cazador bagging twenty-one in a day near Algeciraz; any way those who wish for good Woodcock-shooting had better not try either Andalucia or Morocco, but go to the east of the Mediterranean.

My earliest note of the arrival of a Woodcock about Gibraltar was on the 17th of October, but very few arrive until the middle of November. The latest noticed was on the 8th of March; but I have seen them in Seville market on the 22nd of that month. I obtained near that city a fine white variety, which is now in the Norwich Museum.

I do not like to give second-hand information; but the postmaster at San Roque, Mr. Macrae, an official well known to those who have passed any time at Gibraltar, and upon whose veracity and knowledge of the bird I can depend, told me that once, and only once, he saw at break of day a regular flight, or what the Spaniards would call a "band" of Woodcocks passing south. He described them as being about twenty or thirty in number, but the light was so dim he could not see where they went to.

248. NUMENIUS ARQUATA, Linn. The Common Curlew.

Moorish. Bou-khélal (*Favier*).   *Spanish.* Zarapito.

"This bird is, near Tangier, only a winter resident, which arrives in September and October and leaves during March. They frequent the mouths of rivers and the sea-shore in large numbers, but they are very wild and difficult to get a shot at."—*Favier.*

When at Larache towards the end of April, I observed several Curlews; and a Spaniard who resides there asserted that this bird nests near the town. As Curlews are occasionally seen throughout the summer months on the Spanish side of the Straits, no doubt they are also seen at Larache; but these, I imagine, are birds that do not breed and consequently do not migrate north.

The Curlew is very plentiful near Gibraltar during the winter months, being, perhaps, most frequent in February; but it is very wary, as it is everywhere else in the world that I have met with it. In this species the axillary plumes are white, occasionally marked with brown.

249. NUMENIUS PHÆOPUS, Linn. The Whimbrel.

Favier's notes on this species are the same as on the Curlew, except that he adds "this bird arrives earlier from the north, and though very common, does not remain in the neighbourhood of Tangier for the winter, but passes on further south."

On the Spanish side of the Straits, the Whimbrel is plentiful in autumn and spring up to the end of April, and is occasionally seen in winter; it is, as elsewhere, far less wary and difficult to approach than the Curlew.

In this species the axillaries are white, barred with brown, and the top of the head is also brown.

250. NUMENIUS HUDSONICUS, Lath. The American Whimbrel.

A specimen of this Whimbrel was obtained by Lord Lilford

on the Coto de Doñana on the 3rd of May, 1872 (Ibis, 1873, p. 98). This American species is of the same size as the *N. phæopus*, but can be distinguished by its rufous axillaries.

251. NUMENIUS TENUIROSTRIS, Vieill. The Slender-billed Curlew.

This small Curlew occurs in spring and autumn about the Straits, and is only to be distinguished from the Whimbrel upon close examination, the difference being that the present species has the under wing-coverts and axillary plumes white, the top of the head light-coloured, and the sides of the breast marked with conspicuous pear-shaped spots.

Family GRUIDÆ.

252. GRUS COMMUNIS, Bechst. The Common Crane.

Moorish. Gharnook (*Favier*).   *Spanish.* Grulla.

"This Crane, common in flocks, is found in Morocco only during the winter season, arriving in October and November; they leave for the north in February."—*Favier*.

On the Moorish side of the Straits the common Crane does not appear to remain to nest, as I looked in vain for it in the marshes there during the month of April.

On the Spanish side some thirty to forty pairs breed in the district (comprising many thousand acres) which extends from Tapatanilla along the Laguna de la Janda to Vejer, and thence eastward to Casa Vieja. These birds commence to lay about the last week in April, constructing their nests somewhat like that of the Swan, of sedges, grass, and rushes. The nests vary much in size, some being quite five feet across, others perhaps not much more than eighteen inches: some are deep, and stand high up; others are almost level with the water, in which they are always built. The nest is always placed among sedges or rushes sufficiently short for the bird, when standing up, to be able to see around, and is never built in tall reeds. They are very easy to find, as the old birds never fly direct to the nest, but alight some twenty or thirty yards away and, walking up to it, form regular tracks like a cattle-path; so by following one of these tracks you may be sure of finding the nest: nor do the old birds fly straight away from the nest, but walk off quietly to the end of one of these paths and then take wing. When approached while sitting on the nest, the bird slips off, crouches down, and runs away for some yards.

Mr. Stark watched a pair of Cranes for two or three days from a hill which directly overlooked a marsh where the process of building was being carried on; and he informed me that only one bird worked at a time, the other standing on guard. The nests are never in very close proximity to each other, and never contain more than two eggs, placed side by side so as almost to touch, both the small ends pointing in the same direction. Sometimes the second egg is not laid until two or three days after the first. They differ much in size and shape in different nests; but the pair in a nest are always alike in size, shape, and colour. The latter varies from light buff to an olive-brown, sometimes marked all over with brown and reddish-brown spots, generally thickest at the larger end; but some eggs are almost spotless.

These noble-looking birds are very much harassed during the breeding-time; and being said, I believe correctly, not to lay a second time in the season after the nest has been robbed, they will, I am afraid, soon cease to breed near Casa Vieja, as they have almost done in the marismas of the Guadalquivir, owing to ceaseless persecution. According to what one hears, they used years ago to nest there in great numbers. However, it is the same story everywhere: all wild birds are in Europe certainly decreasing at their breeding-places, owing to egging, drainage, and what is termed civilization; and soon it will come to nothing but Dorking Fowls and domestic Pheasants.

These Andalucian-breeding Cranes are reinforced by the autumn migration, which arrives early in October; and they then form immense bands of from two to three hundred in number, though generally they keep in smaller lots of from five to thirty or forty. Those which do not remain to nest, pass north in March. On the 11th of that month, in 1874,

Mr. Stark and myself had the pleasure of seeing them on passage; and a grand and extraordinary sight it was, as flock after flock passed over at a height of about two hundred yards—some in single line, some in a **V**-shape, others in a **Y**-formation, all from time to time trumpeting loudly. We watched them for about an hour as they passed, during which time we calculated that at least four thousand must have flown by. This was early in the morning. We were obliged to continue our journey; and when we lost sight of the vega of Casa Vieja, over which the Cranes were passing in a due northerly direction, there appeared to be no diminution in their number, and, as my friend remarked, "One would not have believed there were so many Cranes in all Europe." These birds must have crossed the Straits from Africa that morning, the place over which we saw them passing being not twenty miles in a direct line from Tarifa, and a line drawn in the direction from which the birds came would have fallen a little to the west of that town.

Cranes are easily shot in the evening by waiting for them in the swamps where they resort to pass the night. They "flight" earlier than Ducks; and although in the daytime no bird is so wide awake, they are quite stupid in the dusk, flying, if you keep perfectly still, within a few yards. It is, however, a barbarous shame to shoot such a fine and noble bird. Although the Spaniards gladly take them to eat, to my mind their flesh is coarse and worthless; but in India, where they feed much on grain and on rice-stubbles, they are, on the contrary, much sought after for the table. One or two which I shot in the evening at Casa Vieja had been eating beetles and insects, which in winter seems their chief food. They also do a great deal of damage to beans when ripening and to newly sown grain of all descriptions.

253. GRUS VIRGO, Pall. The Demoiselle or Numidian Crane

The only note which Favier has relative to this handsome Crane is that "it is scarce and seldom obtained near Tangier, passing northwards without making any stay, during March, April, and May."

Favier's successor at Tangier evidently considered this species a rare bird; for he asked fifteen dollars (over £3) for a specimen, and at that price it is likely to continue for some time on his hands. He stated that the local name was "Bou-gernan" (father of thistles); but if the bird be as rare as Favier implied, how could it bear a local name?

On the Spanish side I failed to meet with this Crane near Casa Vieja, but strongly suspect that in some seasons it nests there; indeed a pair of Cranes' eggs that were brought to me were so small that I could not refer them to *Grus communis*, but could of course obtain no reliable information about them. Indeed, an egg unidentified is worse than useless to the ornithologist; and unless the collector takes and identifies specimens himself, he had better leave them alone.

In the marismas of the Guadalquivir there is no doubt that in former years the present species used frequently to breed. Specimens are often to be obtained at Seville during March, April, and the early part of May, and again in August. Judging from this, they must nest somewhere a little further north.

Family ARDEIDÆ.

254. ARDEA PURPUREA, Linn. The Purple Heron.

Moorish. Siad el mraj (the hunter of the marsh). *Spanish*. Garza.

"This Heron is, in Morocco, a summer visitant, and nearly as numerous as the common Heron. They pass north in April, returning in September, many remaining in the country to breed, frequenting reed-beds and rushes on the edges of lakes and rivers."—*Favier*.

The Purple Heron, in Andalucia, only remains for the nesting-season; and I never knew an instance of its occurrence in winter. My earliest dates of arrival observed near Gibraltar were the 4th of April 1870, 7th of April 1871, 25th of March 1872, 7th of March 1874. They are extremely abundant and generally easy to get a shot at, being seldom found in the open, but almost always among rushes or swampy jungle, and are very rarely seen to perch on trees. There is, about five miles from Gibraltar, beyond the first river (Guadarranque), on the right of the road to Los Barrios, a leech-preserve, grandly called the " laguna," perhaps two acres in extent and surrounded by poplar trees. This swamp is a dense mass of tall rushes springing up through masses of dead ones, the growth of years past, all so matted and tangled together as to make it very difficult to pass through them, more especially as the water is in places up to one's armpits. This delightful spot is a very favourite nesting-place of the Purple Heron; and there generally used to be three pairs nesting there, also two nests of Marsh-Harriers.

These Herons commence to lay about the 13th of April, as a rule depositing three eggs (rarely four), as the following few instances of nests taken and seen will show:—on the 21st of April two nests—one with four, one with three eggs, all fresh; on the 18th of April two nests—one with one, the other with three fresh eggs; on the 6th of May two nests—one with three fresh eggs, the other with three eggs hard sat-on. The nests, varying much in size and consisting merely of a few dried rushes collected together so as to form a sort of platform just clear of the water, are generally twenty or thirty yards apart. The eggs are light bluish green, similar in colour to those of *Ardea cinerea*.

It is rather remarkable that Purple Herons should generally choose their building-places near to Marsh-Harriers, as the latter repeatedly rob them of their eggs. Many a nest have I seen with nothing but empty shells, the work of the egg-sucking Harrier.

255. ARDEA CINEREA, Linn.   The Common Heron.

Moorish. Aishoush, Bou-ank (*Favier*).   *Spanish.* Garza.

" This species is, in the vicinity of Tangier, both resident and migratory. Those which migrate pass over to Europe during February and March, returning in November and December, being at all seasons plentiful."—*Favier.*

The common Heron visits the neighbourhood of Gibraltar in great numbers during the winter season; and they particularly frequent the district "between the rivers" near Palmones. Mostly departing by March, some few pairs are resident about Casa Vieja; but I never found a nest.

Some of the numerous nests which I saw and supposed to belong to the Purple Heron, might possibly have had the present species for their rightful owner.

The Great White Heron (*Herodias alba*) I saw once at the lakes of Ras-Doura, on the 26th of April, but was unable to get a shot at it.

On the Spanish side I never saw one, or heard of a specimen being obtained. I know the bird well, having shot them both in the Crimea and in India, and could not be well mistaken; and although seeing is believing, seeing only is not sufficient evidence to include a bird in a list.

The Great White Heron has the bill yellow in winter, black when breeding, the legs and feet black; and the total length is about 36–42 inches.

256. HERODIAS GARZETTA, Linn. The Little Egret.

Moorish. Bou-fala, Bou-bliga, Bou-bilira (*Favier*). *Spanish.* Garza blanca.

"This bird is not uncommon near Tangier in small flights when on migration. They pass north in April, returning during November and December; but some remain to breed in the country."—*Favier*.

The Little Egret is the least common of the small Herons in Andalucia, and, as Favier observes, some remain very late, as I have seen and obtained them on the 17th of November. The greater quantity arrive about the middle of April, and linger here and there on their route, gradually passing on to their breeding-places on the borders of the marisma. They nest on trees, in some seasons, near Rocio, but are so molested that they change their ground frequently. When on the wing, and within a short distance, the black legs and bill are very apparent; and this, added to the fact of their being smaller than *H. alba*, serves as a good distinguishing mark for the species. The total length is about 24 inches.

257. ARDEOLA RUSSATA, Wagl. The Buff-backed Heron.

Moorish. Tair el bukkar (the Cow-bird). *Spanish*. Garrapatosa, Purgabueyes.

"This is the most common of the Herons around Tangier, and keeps in small flocks, always following herds of cattle, often sitting on their backs, and chiefly feeding on insects. A small proportion remain during the breeding-season; but the majority pass northwards in February, March, and April, returning late in the year."—*Favier*.

The Buff-backed Heron is very common in low-lying districts in Andalucia, and some are resident; but they are very irregular in their movements, and chiefly noticed, while passing, during March and April, as they always attend cattle when in wet marshy ground. The Spanish herdsmen naturally object to have them molested, especially as there is a story of a sporting Briton from Gibraltar having shot one as it sat on a cow's back—a story which I am afraid is founded on fact, and only shows what the Englishman is capable of.

The local names of this Heron all originate from its habit of attending cattle and freeing them from parasites—*Garrapatosa* from *garrapata*, a tick or louse; *Purgabueyes*, cattle-cleaner or purifier.

A male bird, which had been kept alive for about four years in the patio of the Fonda de Europa, at Seville, during the first week in April (his fifth spring, as far as I could ascertain) began to change the colour of the legs and the basal half of both mandibles to a pinkish red; the irides also changed to beautiful rich pink colour, with a very slight golden ring round the black pupil. This change was quite completed before the bird had fully assumed the buff-coloured back, which is the mark of the breeding-dress.

A female, in confinement with the above mentioned, has laid many eggs of a very pale bluish-white colour, showing a greenish tint inside when held to the light.

These captive Herons were quite masters of the various Kites and Buzzards confined in the same patio, and ceaselessly wandered around, hunting flies, which they caught when settled on the walls or ground, never attempting to take them on the wing; but, poising the head two or three times, as a man would a dart before throwing it, they never missed their aim.

This species breeds in the marismas; but I have no personal knowledge of its nesting-habits. In the winter the adults are white, excepting on the crown of the head, which is marked with buff.

Length about 20 inches; larger than *Ardeola comata*, smaller than *Herodias garzetta*.

258. ARDEOLA COMATA (Pall.). The Squacco Heron.

Moorish. Aishus (*Favier*; but he applies this name to all the Herons). *Spanish.* Garza canaria (from its colour).

"This species is nearly as common around Tangier as *Ardeola russata*, occurring in small flocks during migration. Some remain in the country to breed, nesting on the ground among sedges, laying in May and June five eggs, which are more oval in shape than those of *A. purpurea*, but of the same colour."—*Favier*.

I found this Heron in great numbers about the swamps of Ras-Doura towards the end of April; and they were, there, by far the most common species of *Ardeidæ*.

On the Spanish side the Squacco Heron is entirely migratory, arriving during the month of April. They are common in the marisma of the Guadalquivir; but I never observed any near Gibraltar, nor did I ever see them following cattle, like the preceding species. They nest late in the season; but I regret to be unable to give any personal information as to their breeding-habits.

This species (beautiful as all the family are) is, to my mind, by far the most handsome and elegant of all the European Herons. It is, at the same time, the smallest of them, and further to be distinguished by the crest, which varies in length and consists of from eight to ten pointed white feathers bordered at the sides with black.

The total length is about 19 inches.

259. ARDETTA MINUTA (Linn.). The Little Bittern.

"This species is, near Tangier, the most scarce of the *Ardeidæ*, being not often met with, and then always either alone or in company with *Ardeola comata*. They arrive and pass on north in April, and return during August to winter further south."—*Favier*.

The Little Bittern is, in Andalucia, entirely migratory, arriving late in April. Considerable numbers nest among rushes and sedges. They are late-breeders, nesting early in June, and laying as many as six white eggs. I have no exact date of the autumnal migration; but they are all gone by October.

260. NYCTICORAX GRISEUS (Linn.). The Night-Heron.

Favier says:—"This species is common near Tangier when on migration, passing in small lots, which frequent wooded spots close to lakes and rivers." I saw the Night-Heron near Larache in April and near Tetuan at the end of March.

In Andalucia they are entirely migratory, chiefly arriving in April; but I have no date of their autumnal departure, and never observed any very near to Gibraltar. About the district of Seville they are common, nesting in companies on trees on the Rocina near Rocio and on the banks of rivers— like the other smaller Herons, breeding rather late.

The Night-Heron, as its name implies, is a nocturnal-feeding bird, frequenting trees by day, and if disturbed usually flying from one tree to another; but I have scarcely ever seen them on the move by day, unless frightened up.

The immature birds, in their brownish spotted plumage, are, but for their arboreal habits, at a little distance very liable to be mistaken for the Bittern (*Botaurus stellaris*).

261. BOTAURUS STELLARIS (Linn.). The Common Bittern.

Moorish. Sebar el Mraj (Lion of the marsh). *Spanish.* Pajaro toro (Bull-bird), Guia de las Gallinetas (Guide of the Woodcocks).

The Bittern, according to Favier, " winters in Morocco, and is seen in abundance on passage, arriving during August and September, and leaving in February. They are found in pairs and in small lots, frequenting rushes and reed-beds."

This bird, however, breeds as far south as the neighbourhood of Rabat, whence I have seen the eggs. On the north side of the Straits they sometimes nest at Casa Vieja, at the Laguna de la Janda, and in the Soto Torero near Vejer, and also in the marshes of Rocio, near the Coto Doñana. Nesting about the middle of May, they lay four or five pale brown-coloured eggs, placed in the midst of thick rushes. I was unable personally to find a nest, but had several eggs brought to me, and have often heard them calling in the daytime—a peculiar, booming, unmistakable cry, whence, in almost all countries, their local name is derived.

More abundant in the winter months, they arrive in the end of October, and in some places are at times quite numerous wherever there are rushes and sedges; and I have occasionally known them shot in the sotos of the Cork-wood.

They are dull and sluggish in habit, and it is not until nearly trodden on that they will rise; but on one occasion I remember finding several in some rather open marsh, and they flew up one by one far out of shot, seeking refuge in the nearest thick reed-bed. Though often flushed among sallows and bushes in the Soto Malabrigo, near Casa Vieja, I never once saw the Bittern perch on bushes or trees there or in any other country.

Family CICONIIDÆ.

262. CICONIA ALBA, L. The White Stork.

Moorish. Belarcjh. *Spanish.* Cigüeña.

"This Stork is seen on migration in vast numbers around

Tangier, passing to Europe during January and February, some of the birds terminating their journey by remaining to breed in Morocco. These are the first to depart south, returning again year after year to the same places, and apparently by the same route as that taken in their gradual departure.

"Some large flights pass on without stopping; those which migrate in August rest awhile on their way south; so during the autumnal migration (which lasts, like the spring, for about a month, the latter half of August and the first part of September) this species is extremely numerous and seen around the environs of Tangier in all directions; they are very tame, and often follow close behind the plough.

"The superstition which shelters this bird from molestation by the natives has been mentioned in my notes on the Swallow; but it may be added that some of the Arabs believe that the Storks originate from a wicked Kadi and his family, who, as a punishment for their great cruelty, were all changed into these birds, and that these *misérables* humble themselves to appease Allah, and, in the hope of some day regaining their original human form, pray without ceasing day and night, and, whenever they rest, prostrate themselves and clack their bills."—*Favier*.

The White Stork, owing to the protection it everywhere receives, is much more abundant in Morocco than in Andalucia, although it is plentiful in some level districts in the latter country, being most common in the marismas and in the vicinity of Seville, nesting on some of the churches in that city. On the African side of the Straits, in many situations they breed on trees, generally in colonies, as well as on houses, but usually near villages; and almost every Moorish hovel has its Stork's nest on the top, a pile of sticks lined with grass and palmetto-fibre. It usually contains four white eggs, which are very rarely marked with pink blotches; these are sometimes laid as early as the 25th of March, and are very good eating, either hot or cold. When boiled hard, they have the white clear, as with Peewit's or "Plover's" eggs, the yolk being of very rich reddish yellow.

The White Stork is rather irregular as to the time of

nesting; for I found in Morocco on the same day (the 25th of April) young birds, eggs, and unfinished nests; and, to show how varied is their time of migration, I saw on that day a flight of about a hundred, flying northwards at an immense height. As they passed over the "Storkery" they lowered themselves to within a hundred yards or so of the nests, and after wheeling round a few minutes, as if to see how affairs were going on, they worked up in a gyrating flight to their original elevation, and continued their northerly journey, doubtless to the great delight of the resident Storks, who were in a great state of perturbation and disturbance at the appearance of their brethren. I may here remark that Storks usually migrate in large flocks at a great height, with a gyrating flight. The earliest date of their arrival that I noticed near Gibraltar was on the 11th of January; and they nearly all leave by the end of September. Feeding on insects of all kinds, mice, snakes, and other reptiles, they are most useful birds, and certainly deserve the protection and encouragement which they receive in Morocco, where they are in consequence excessively tame. Their grotesque actions when nesting, and their habit of continually clacking their bills together, making a noise like a rattle, render them very amusing to watch. I was informed by a Frenchman who had passed two years in the city of Morocco, that there, as well as at Fez and some other large towns in the Moorish Empire, there is a regular Storks' hospital, and that should one be in any way injured, or fall from the nest, it is sent to this institution, or, rather, enclosure, which is kept up by subscription from wealthy Moors, who consider the Stork a sacred bird. I merely mention this story to draw attention to the subject in case of any future ornithologists visiting these cities; and were not my informant worthy of credence, should have omitted noticing it.

263. CICONIA NIGRA, Linn. The Black Stork.

Spanish. Cigüeña negra.

"This species is much less common in the vicinity of Tangier than the White Stork (*C. alba*). They are seen crossing the Straits during the months of February, March,

April, and May, returning in November to pass further south. During their passage, they keep in pairs and in small lots, frequenting much the same ground as the Crane (*G. cinerea*)."
—*Favier*.

Favier also gives a Moorish name, thus written, *Gerïnga*; but it is probably incorrect. I repeat it in case it should be all right. From its shy and wild and, as far as my observations go, solitary habits, the Black Stork on both sides of the Straits appears to be much less common than perhaps is really the case. I saw one near Tangier in October 1869, and another on the 26th of April 1871, and have seen several specimens obtained in the neighbourhood. Near Gibraltar I saw one on the 22nd of February, another on the 11th of January 1872, near Seville, and obtained a specimen from there on the 18th of November, 1870. These were the only instances when I personally noticed it.

Family IBIDIDÆ.

264. IBIS FALCINELLUS, Linn. The Glossy Ibis.

Moorish. Maiza (*Favier*). *Spanish.* Morito.

" This bird occurs near Tangier on passage, returning to pass the winter further south. Some must remain to nest in the country; for they are frequently met with during May, June, and July."—*Favier*.

We saw great flocks of the Glossy Ibis at the lakes of Ras Doura towards the end of April; but they were very wary, as they are in Andalucia, whereas I remember in India one used to walk up to within thirty yards of them.

When flying they much resemble the Spoonbill in their manner of flight. They nest in Morocco, as I have seen eggs obtained in the country; and I am well informed that in wet seasons they breed in the Soto Torero, near Vejer, and also in the marismas of the Guadalquivir; but I have no personal knowledge of their nesting-habits. The eggs are of a uniform pale bluish green colour. Near Gibraltar I have only noticed this Ibis when passing late in April and in May. A female shot at the First River on the 31st of May, had the gizzard full of minute shrimps; four eggs in the ovary were slightly enlarged, which tends to confirm what I have heard, that they are late breeders.

I may as well here mention that Favier includes in his MS., without any description, another species of Ibis as having once been obtained by him near Tangier. He calls it "*Ibis calva*;" but it could hardly have been that South-African species, and was much more probably *Geronticus comatus*, Ehrenberg, the only information concerning which bird that I have any access to being given by Dr. Tristram in 'The Ibis' for 1860 (p. 78), where he describes it as a mountain-haunting species.

265. PLATALEA LEUCORODIA, Linn. The White or Common Spoonbill.

Moorish. Bou-ka-kaba (*Favier*). *Spanish.* Espatula, Paleton, Paleta, Patera, Pilato.

"This species occurs near Tangier when on passage. They migrate north in March, April, and May, returning during October, and are never observed in winter."—*Favier*.

I saw many Spoonbills in April at the lake of Masharal-haddar, near Larache; and they then appeared to be on migration. The earliest occurrence of this species in spring near Gibraltar that I know of was one shot on the 9th of April, at the First River; and the latest I saw was a single bird wading about the river Barbate, near Casa Vieja, on the 20th of November. They were common in the marismas in flocks in May; in some wet seasons they nest there, and also in the Soto Torero, near Vejer, where, sad to relate, a Spaniard, in 1873, took upwards of seventy eggs early in May. He took most of these eggs into Gibraltar, to some collectors who were there at that time; and next year he described to me the nests as merely made of a few sedges, and placed close (*junto*) together, each containing four eggs. The season of 1874 was very dry, and no Spoonbills appeared there; indeed, had it been wet, probably after being so robbed, the poor birds would not have nested again in that spot. What have not collectors to answer for?

Order ANSERES.

Family PHŒNICOPTERIDÆ.

266. PHŒNICOPTERUS ANTIQUORUM, Temm.   Flamingo.

Moorish. Nihof.   *Spanish*. Flamenco.

Favier says:—" The Flamingo, near Tangier, passes northwards in April, May, and June, returning in August up to as late as December. The females are the first to arrive during the autumn migration. The males rejoin their mates in November, accompanied by the young of the previous year; the young of the year are never seen here. They are met with in large flocks on the lakes, always staying in the water, though they never swim about, and are very wary and difficult to approach. The only month in which they are entirely absent is July. Their temporary absence during other months is regulated by the quantity of water in the lakes; and as one month is not sufficient time for them to lay and hatch their eggs, they ought to nest not far from Tangier: indeed an old chasseur, worthy of belief, informed me that he had shot one which, when it fell, dropped an egg in the water."

The movements of the Flamingo are certainly very irregular and perplexing, and, no doubt, influenced by the amount of water in the brackish lagoons which they frequent. Most of these lagoons, being formed by rain-water, are brackish from the salt contained in the earth, and in very dry seasons hold hardly any water.

In very wet seasons the birds breed in the marismas of the Guadalquivir, and are said to nest very late (about June). The exact manner of nesting is at present unknown to ornithologists; and he who first finds and describes it will have " a feather in his cap."

The eggs which I have seen are elongated and of a white colour, with a chalky surface.

Flights of Flamingoes are frequently seen passing near Gibraltar as early as the 4th of February and as late as the

1st of May; and they again appear in September, when immature birds are met with. I have seen flocks of thousands in the marisma near the Isla Menor, and, by the aid of a stalking-horse, managed to shoot five at a shot. Usually they are extremely wild and shy, except during actual passage, when they alight to rest at the mouths of rivers.

The note is not unlike that of the Grey Lag Goose (*Anser cinereus*); and more than once at night I have mistaken the sound for that of these Geese.

267. CYGNUS MUSICUS, Bechst. The Hooper or Whistling Swan.

Spanish. Cisne.

This is the only species of Swan which I was able to identify in Andalucia, having examined one specimen shot on the Guadalquivir below Seville, where they are said in some winters to be common.

The Hooper, in its adult plumage, has the bill yellow, with black tip, edges, and nostrils.

268. CYGNUS OLOR, Linn. Mute Swan.

Favier says:—"This Swan is tolerably numerous, and seen flying over near Tangier in small flights, rarely remaining in the vicinity; but they did stay in 1845 and 1849. They pass south in December, returning in April."

The distinguishing mark of the adult birds is the black tubercle on the upper part of the orange-coloured bill.

Family ANATIDÆ.

269. ANSER CINEREUS, Meyer. The Grey Lag Goose.

Moorish. Wiz. *Spanish.* Ganso, Anser.

The above names equally apply to *A. segetum*.

Favier's notes are the same for both this and the following species, viz.:—"This Goose, which the Arabs confound with *Anser segetum*, is as numerous as that bird near Tangier, arriving during November and December. They retire north in March, seldom making any stay near Tangier; they pass on to the large lakes and rivers."

On the Spanish side of the Straits the Grey Lag Goose is found in winter at the Laguna de la Janda and in the various lagoons of the marismas of the Guadalquivir in enormous numbers. They generally arrive at the former place about the 20th of November, the earliest that I noticed in two consecutive years being on the 8th of November and the 25th of October. Commencing their departure about the 14th of February, they are all gone by the first week in March, and seem for the most part to migrate by day. Although, like Ducks, they "flight" at night (though, as a rule, rather later in the evening and later in the morning), they affect particular favourite spots and pools without any apparent reason for their likes and dislikes, some places never being frequented by them.

The Grey Lag Goose can always be easily distinguished at some distance on the wing by the ash-grey of the shoulders, which colour, when they are flying, is very apparent. They make a creaking noise, caused by the stiff primaries, somewhat resembling the rattling together of dry reeds, which can be heard only when they pass very close; and a very joyous sound it is for the sportsman to hear.

270. ANSER SEGETUM, Gm. The Bean-Goose.

On the Spanish side of the Straits this species is much less numerous than the Grey Lag Goose; and it was some time before I could succeed in obtaining a specimen for identification. As far as my experience goes, I should say the present species occurs in the proportion of one to every two hundred of the Grey Lag; but as Favier considers both kinds equally common in Morocco, perhaps in some seasons the present species may be more abundant than in others.

The White-fronted Goose (*Anser albifrons*) and the Pink-footed Goose (*A. brachyrhynchus*) I never met with on either side of the Straits. As it is possible they may occasionally occur, I give the distinctive marks and relative measurements of the four common European Geese, in the hopes that some one may meet with the latter species in Andalucia:—

	Length of a male.	Colour of feet and legs.	Colour of nail of bill.
	in.		
Grey Lag Goose (*Anser ferus*)	35	Dull flesh-colour	White.
Bean-Goose (*A. segetum*)	34	Orange	Black.
Pink-footed Goose (*A. brachyrhynchus*)	28	Pink	Black.
White-fronted Goose (*A. albifrons*)	22	Orange	White.

271. BERNICLA LEUCOPSIS (Bechst.). The Bernicle Goose.

A single specimen of this Goose, obtained near Seville several years ago, was in the possession of the landlord of the Fonda de Europa, possibly an escaped bird from San Lucar. This is the only instance in which I have heard of its occurrence in Andalucia.

The Bernicle Goose has the top of the head and all the neck black, the forehead, cheeks, and chin white; and the total length is about 25 inches.

272. TADORNA VULPANSER, Fleming. The Common Sheldrake.

Moorish. Bou-ha-baïda (*Favier*). *Spanish*. Pato tarro.

" This species is not regular in its appearance near Tangier, and occurs between November and February."—*Favier*.

I never personally met with this Sheldrake on either side of the Straits. A regular maritime or sea-coast bird, they are found on the coast near the mouth of the Guadalquivir, whence I have seen specimens; and there is no doubt that they breed there.

This bird has the head and neck green. The males resemble the females in plumage, but are not quite so bright in colour.

The total length is from 24 to 26 inches.

273. TADORNA RUTILA (Pall.). The Ruddy Sheldrake.

Moorish. Bou-ha (*Favier*). *Spanish.* Pato tarro.

"This species is resident at no great distance from Tangier; and others are migratory, crossing to Europe during April and May, returning in September and October. In the immediate vicinity of Tangier it is scarce and only observed in small lots on the lakes and large rivers. Often they entirely, though irregularly, disappear for months at a time, probably going to marshes not very far off. The months during which they are usually absent are February, March, and June."—*Favier.*

The Ruddy Sheldrake, best known to Anglo-Indians as the Brahminy Duck, is quite an inland bird, though sometimes frequenting salt lagoons, and is much more frequent on the African than on the Spanish side of the Straits, where I never met with it alive, though I have seen a few in Seville market in spring and the end of autumn. They are said to nest near the mouth of the Guadalquivir; and there can be no doubt but they do breed somewhere north of the Straits. They nest in holes in cliffs and rocks; and an account is given of one nest by Mr. Salvin, in 'The Ibis' for 1859 (p. 362). In Morocco they are very wary; and I recollect, in company with a brother officer, trying in vain to approach three which we saw near Vincent's Farm, Sharf el Akab, in October: but at last they luckily pitched close to a horse feeding in the marsh; so we stalked up behind the animal to within twenty-five yards of the Ducks, bagging all three.

I have repeatedly seen them exposed for sale in Tangier market, with their throats slit in Mahometan fashion.

This Sheldrake has the head of a buff colour, darkening on the neck to a rufous brown, the adult male having a black ring at the base of the neck.

The total length is about 25 inches.

274. SPATULA CLYPEATA (Linn.). The Shoveller.

Moorish. Bou-slafa (father of the bowl). *Spanish.* Paletone, Sardinero.

"This Duck is, in some winters, common near Tangier,

arriving during September and October, leaving for the north in February and March."—*Favier*.

The Shoveller is met with in considerable numbers on the Spanish side of the Straits. They mostly arrive around Casa Vieja and on the Laguna de Janda in October; and I have known of their occurrence there in August, but have no certain knowledge of their nesting in the neighbourhood; indeed I never saw any later than the end of April.

This (the only species of the genus to be found in Europe), owing to the shape of the bill, which is dilated broadly at the end, cannot be well mistaken for any other Duck. The alar speculum is metallic green.

275. CHAULELASMUS STREPERUS (Linn.). The Gadwall.

"This species is as scarce near Tangier as *Fuligula rufina*, the Red-Crested Pochard; and their appearance, which takes place between February and March, is irregular and uncertain."—*Favier*.

The Gadwall, on the Spanish side of the Straits, cannot be termed a common bird. I only met with it on a few occasions: one shot at flight on the 26th of November 1869, another shot at flight on the 22nd of December 1871 at Tapatanilla, and three others killed there in February 1874, are my only personal experiences of it; but I have seen it in Seville market in February and March.

Lord Lilford informed me that he saw ten or twelve Gadwall at the lakes of Santa Olaya, in the Coto de Doñana, in the early part of May, and considered that they were breeding, although he was unable to discover a nest. He also informed me that the local name there was "Frisa"—a word which signifies coarse cloth or frieze.

This Duck, with its plain plumage, can be easily distinguished by the colourless or *white* alar speculum, the outer webs of the secondaries being white. The male has the point of the wing and the small coverts bright chestnut; the legs are a dirty yellow.

The total length is from 19 to 21 inches.

276. ANAS BOSCHAS, Linn. The Wild Duck.

Moorish. Zerak el ras (Blue head). *Spanish.* Pato real.

"The Wild Duck frequents the vicinity of Tangier throughout the year. Those which are not resident cross to Europe during March and April, returning in November and December. Those which remain to nest begin to lay during the month of February; and eggs may be occasionally found as late as the beginning of June. Tame Ducks are called *Bourk* by the Arabs."—*Favier.*

In Andalucia the present species is abundant in winter; and a considerable number remain for the breeding-season, hatching by about the 25th of April; but they are so molested by egging that it is a wonder that any young are brought up; and in addition to this, they are shot at all seasons.

The Moorish local name given above is not appropriate, the head of the Mallard being green.

In this species the alar speculum is metallic blue.

277. ANAS ANGUSTIROSTRIS, Ménétr. The Marbled Duck.

Moorish. Shihib (*Favier*). *Andalucian.* Ruhilla.

This Duck, on both sides of the Straits, appears in spring, to remain only for the breeding-season, and is exceedingly abundant in Morocco, where, at the lakes of Ras-Dowra in April, I saw flocks numbering many hundreds; and they are frequently seen exposed for sale in Tangier market.

Favier states that they arrive during March and April, departing in October, and that after the common Teal they rank as the most common Duck in the country.

On the Spanish side I heard of three being seen at the end of February, and saw six or seven myself on the 23rd of March; but the majority do not appear until late in April, though I have noticed them on the sea near Gibraltar early in that month. As a rule, they all leave by the end of September; but of course stragglers remain later.

The Marbled Duck breeds during the last week in May, nesting in patches of rushes. The nest is like that of a Teal, containing a good deal of the down from the breast of the female; and eleven eggs appear to be the usual complement

The latter much resemble those of the common Teal, being of a yellowish-white colour. Favier states that they also nest in rushes during May and June, and that incubation lasts from twenty-five to twenty-seven days.

I was unable to find the Marbled Duck near Casa Vieja or about the Laguna de la Janda, nor could I ascertain that it is known there; but in the marismas of the Guadalquivir, especially near the Coto del Rey, it is not uncommon.

In flight, the Marbled Duck somewhat resembles the female Pintail; but it is more of a Teal, as Lord Lilford observes. I found them wary and difficult to approach; but in the dusk they "flight" very low, and by watching the direction taken by them for one night you may on the next evening be tolerably certain of shooting a good many; and they are excellent eating.

The males resemble the females in their sombre dress, being of a dull brown colour, marbled with light grey-brown, not having a bright feather in the plumage; the alar speculum is pale creamy brown. The bill is narrow; hence the specific name.

Total length about 14·5 inches, tarsus 1·2.

278. QUERQUEDULA CIRCIA (Linn.). The Garganey Teal.

"This Teal appears irregularly near Tangier, only on migration, and does not occur every year. They arrive during February and March, passing on to the north, and are seen returning south in September."—*Favier*.

The Garganey Teal seems to be equally irregular in its appearance in Andalucia, as I only saw one in Seville market in March 1869, and did not again meet with it till March 1874, when they were for a few days not uncommon about Casa Vieja and near Seville. Lord Lilford informs me that he saw a pair at the Laguna de Santa Olaya, near Rocio, in May 1872—which would render it probable that they sometimes breed in Andalucia.

The adult male has the wing-coverts bluish grey; in the female they are ashy grey. The alar speculum is green.

Total length from 14 to 16 inches.

279. QUERQUEDULA CRECCA (Linn.). The Common Teal.

Moorish. Frifar *(Favier).* *Spanish.* Sarceta.

"This species is abundant near Tangier during winter, passing north in February and March, returning in September and October."—*Favier.*

The common Teal is very numerous on the Spanish side, and, from its haunting small streams and marshes, is more easily shot than any other of the Ducks. Their numbers vary considerably; and in some seasons they are much more abundant than in others. They chiefly arrive during October, leaving for the north in March, and have been known, though very rarely, to remain to breed near Casa Vieja. Lord Lilford observed a single bird at Santa Olaya in May 1872.

Total length 12·5 to 14·5 inches.

Alar speculum green.

280. DAFILA ACUTA (Linn.). The Pintail.

Moorish. Bou-ze-boula *(Favier).* *Spanish.* Pato rabudo, Pato careto.

"This Duck is, during the winter season, nearly as plentiful in the vicinity of Tangier as the Wild Duck (*Anas boschas*). They arrive during September and October, and leave in April and May."—*Favier.*

On the Spanish side I never found the Pintail before the month of November; it is exceedingly abundant during the winter months on the Laguna de la Janda and other large open pieces of water, and is in consequence very difficult to shoot by day without the aid of a stalking-horse. These Ducks mostly depart during March, some lingering on later into the month of April, the latest that I myself observed any being on the 5th of that month.

In this species the tail is cuneiform, consisting of sixteen feathers, the central ones much elongated in the male, very slightly so in the female; the alar speculum is green, glossed with a copper colour.

281. MARECA PENELOPE (Linn.). The Widgeon.

Moorish. Bou-kha-saiwa. *Spanish.* Silbon (Whistler), Pato franciscano.

"This species is the most abundant of all the Ducks near Tangier, being found in large flocks throughout the winter months. They commence to arrive in August and September, and leave during March and April."—*Favier.*

Exactly the same may be said of the Widgeon on the Spanish side of the Straits, except that I never saw any so early as Favier mentions. They commence to arrive early in October; but the greater number do not appear until November; and they are then by far the most common of the *Anatidæ*, in some winters swarming in thousands on the Laguna de la Janda. Their departure for the north begins about the end of March; but a few linger on throughout the whole of April.

In this species the tail consists of fourteen feathers; the alar speculum is glossy green.

282. FULIGULA RUFINA. The Red-Crested Pochard.

"This Duck is accidentally met with around Tangier, but is a very rare species. I only obtained two—one in 1835, the other in 1849."—*Favier.*

I never met with this Pochard on either side of the Straits, and have seen but one specimen said to be Andalucian. A more eastern species (frequenting still, deep waters, and seen rarely on rivers), it is of more common occurrence in the south-east of Spain; and Lord Lilford mentions it as common on the Albufera near Valencia, where it used to breed.

The adult male is a very handsome bird, having the head and upper part of the neck a rich reddish chestnut-colour, the feathers at the top forming a distinct comb-shaped crest; speculum white; the bill, legs, and feet are bright vermilion-red; irides red. The female has a more sombre-coloured dress, and wants the crest, having the *cheeks, throat, and sides of the neck,* as well as the speculum, *greyish white;* the soft parts reddish brown.

Total length about 22 inches.

283. FULIGULA FERINA (Linn.). The Common or Red-headed Pochard.

Spanish. Cabezon.

"This species arrives during October to remain in Morocco for the winter, departing for the north in April and May."—*Favier.*

I found the Pochard common about the lakes near Tetuan, and shot one there as late as on the 30th of March. On the Spanish side of the Straits I have rarely seen this Duck near Gibraltar, and then only in winter; but it is more abundant in the marismas below Seville; and at times a good many are to be seen at the Laguna de la Janda. The Andalucian lagoons, however, being mostly very shallow and void of weeds, are not suited to the habits of this diving Duck.

The adult male has red, the female brown, irides. The speculum is ashy grey; tail-feathers fourteen in number.

Length about 19½ inches.

FULIGULA MARILA (Linn.). The Scaup Duck.

This Duck is a rare visitant in the Straits, but has occurred in the Bay of Gibraltar in December. As it is a coast-frequenting bird, it is not liable to much notice and seldom likely to appear in the markets. Irides yellow.

Total length 19 to 20 inches.

285. FULIGULA CRISTATA (Linn.). The Tufted Duck.

"This species is in some years very abundant near Tangier, arriving here for the winter in November, and returning north during February. In some seasons they are not to be met with, but were common in the years 1845, 1846, 1849, 1850, 1858, and 1861."—*Favier.*

The Tufted Duck is sometimes plentiful in winter on the Laguna de la Janda, and is well known in the marismas. I have occasionally seen them in the Bay of Gibraltar.

The head of the adult male is crested and of a purplish black colour. The female has no crest, and the head is dark brown; irides yellow.

Total length 16 to 17 inches.

286. NYROCA FERRUGINEA (Gm.). The White-eyed Pochard or Ferruginous Duck.

Moorish. Ziriguil (*Favier*).   *Andalucian.* Negrete.

The White-eyed Pochard may be considered, like the Marbled Duck, a summer resident on both sides of the Straits, and is by far the most abundant in Morocco.

I saw many hundreds at the lakes of Ras-Dowra towards the end of April, being even then in large flocks. We shot some at flight in the evening at the same time as *Anas angustirostris*; but the two species did not fly together. There were also a few of the White-eyed Ducks about the lake of Esmir at the end of March. Favier writes of the present species, that it is "abundant near Tangier, arriving from the south during May and departing in November and December, totally disappearing for a time in winter. They are most abundant at Ras Dowra, breeding in June and July, the incubation lasting thirty days."

On the Spanish side, the White-eyed Duck is common during the breeding-season in some parts of the marismas, and commences to nest about the end of April. Lord Lilford obtained a nest in May 1872, in the Coto de Doñana, composed of dead dry water-plants, flags, &c., lined with thick brownish white down and a few white feathers. It was placed at a short distance from the water, in high rushes, and contained nine eggs. Although they generally pass south early in autumn, I once saw and shot a single bird as late as the 6th of December, which, albeit in fair condition, from its excessive tameness, was probably from some cause incapable of migration.

I have always found this Duck, like its allies, *F. rufina* and *F. ferina*, frequenting deep still weedy water rather than shallow open places; and the flesh of the present species is not only, like theirs, excellent eating, but far surpasses either in that respect.

The irides are white, excepting in the young birds, which have them reddish brown; the speculum is white.

Total length from 14 to 16 inches.

287. GLAUCION CLANGULA (Linn.). The Golden-eye.

All to be stated regarding this northern species is that it rarely occurs about the Straits in winter. Irides golden yellow.

Total length of male 19 inches.

288. ERISMATURA LEUCOCEPHALA (Scop.). The White-headed Duck.

" This species occurs near Tangier on passage, passing north during April and returning to winter further south in October. Some of the spring migrants remain in the country to breed in June, laying as many as ten pure white eggs, with a rough granulated surface. This Duck is not at all regular in its appearance, but in some seasons is quite common."—*Favier*.

The White-headed Duck is found chiefly on the coast and on large lakes, and occurs near Cadiz, no doubt breeding in the country, though I never personally met with it on the Spanish side of the Straits.

The white face of the adults, and the long pointed stiff black tail-feathers, of this species will sufficiently distinguish it from any other of the European *Anatidæ*.

Total length 17 to 18 inches.

289. ŒDEMIA NIGRA (Linn.). The Common Scoter.

Moorish. Bourk-el-behar (Sea-Duck).

Favier states that the Scoter is " found in abundance near Tangier, arriving sometimes as early as August, retiring northwards in April."

I found this Duck in some seasons very common about the Straits, especially after rough weather in Gibraltar Bay; but they do not appear except in small lots. The earliest noticed was on the 12th of November; the latest on the 12th of March, 1872.

The males are entirely black, except the upper part or ridge of the bill, which is yellow; the females are of a blackish brown colour. The immature birds have the cheeks, side, and front of the neck greyish white, with white spots on the underparts.

Total length about 22 inches.

The Velvet Scoter (*O. fusca*) is most probably an occasional winter straggler in the Straits, and should be looked for. The adult male has reddish legs and orange bill; the alar speculum in both sexes is *white*, which will serve to distinguish the species.

Length about 23 inches.

Family MERGIDÆ.

290. MERGUS ALBELLUS (Linn.). The Smew.

The Smew occurs in some seasons about the Straits in immature plumage; but I never heard of an adult of any of the *Mergidæ* being obtained. The immature males of this and the next two species can be at once distinguished by the trachea, which in the males has more or less bony enlargement at the base, while in the females it is uniform in size throughout the entire length.

The Smew is the smallest of the family, and has sixteen tail-feathers. It measures from 14 to 17·5 inches in length, females being the smallest.

291. MERGUS SERRATOR, Linn. The Red-breasted Merganser.

This species is not mentioned in Favier's MS., but is found in some winters in considerable numbers in the Bay of Gibraltar, and is generally seen during December and January. I never, however, met with an adult male.

Total length about 21 inches.

292. MERGUS MERGANSER, Linn. The Goosander.

The Goosander is recorded by Favier as having been once obtained by him near Tangier in October 1862. I saw another which had been found dead on the shore near that town during the winter of 1869–70, the only instance in which I met with the species.

The Goosander may be distinguished from the other European *Mergidæ* by its large size, measuring from $25\frac{1}{2}$ to $26\frac{1}{2}$ inches in length.

Family PELECANIDÆ.

293. PHALACROCORAX CARBO, Linn. The Common Cormorant.

Moorish. Gharrad (*Favier*). *Spanish*. Cuervo marino.

"The Cormorant is found near Tangier from December to February, and frequents the coast, lakes, and rivers, where it is not uncommon."—*Favier*.

The above remarks equally apply to this bird on the Andalucian side. I never saw it in summer.

Total length 33 to 36 inches.

294. PHALACROCORAX GRACULUS, Linn. The Shag.

Favier includes this species in his list as *P. Desmarestii*, stating that it "is rare near Tangier, but found during the whole year."

I found the Shag, or Green Cormorant, very common in the Straits. They nest at the island of Peregil, under Apes' Hill on the African coast. I brought home one specimen for identification shot close to Gibraltar.

The adults are of a general green colour, the immature birds being of a brownish hue.

Total length about 27 inches.

295. SULA BASSANA, L. The Gannet or Solan Goose.

Moorish. Bou-grana (Father of frogs) (*Favier*). *Spanish*. Alcatraz; but this name is often applied to any large Gull.

Favier merely remarks of this bird that it "arrives in October and leaves during March, not being very numerous." I always, however, during the winter season, saw great numbers of Gannets in the Straits, particularly close to Gibraltar, where, according to the wind, they might be noticed fishing on the leeward side of the Rock; and many a time have I watched them darting down from a considerable height on their prey, often disappearing quite under the water. On the wing, to an inexperienced observer, they appear like a large Gull. The immature birds in their dull spotted dress, perhaps through not attracting so much notice, seem to be less in number than the more conspicuous white adults with their black primaries. The earliest dates on which I observed this species near Gibraltar were on the 11th of November, 1870, and the 12th of October, 1871, the latest being on the 28th of March, 1870, and the 22nd of March, 1871, and the 28th of March, 1872. I noticed many on the last date; but that spring was remarkable for the late stay of several northern-breeding species.

Family LARIDÆ.
Subfam. STERNINÆ.

296. STERNA CASPIA, Pall. The Caspian Tern.

Spanish for all Terns, Golondrina de mar.

This large Tern is stated by Favier to be "very rare near Tangier," he having only obtained a single specimen (in February 1844), which occurrence I can supplement by one which occurred in the winter of 1869.

On the Spanish coast I did not meet with it; and indeed it is only an accidental straggler so far west.

The large size of this Tern, the red bill, and black legs and feet sufficiently distinguish it.

The total length is from 19 to 21 inches, the males, as in the other Terns, being the largest.

297. STERNA ANGLICA, Montagu. The Gull-billed Tern.

Strangely enough, no mention is made of this Tern in Favier's MS.; but I found it in great numbers about the lakes of Ras Dowra towards the end of April. As far as I could ascertain from the Arabs, they said that these birds remained in the neighbourhood and bred a little later on in the season. Essentially a freshwater or marsh-frequenting species, I never noticed the Gull-billed Tern on the sea-coast. Some of those I shot had been feeding on green frogs; their note, loud and frequently repeated, is (as near as I can render it) *kŭh-wŭk, kŭk-wŭk*. I never noticed the present species about Gibraltar; but it occurs in the marshes of the Guadalquivir towards San Lucar, and no doubt breeds somewhere there; but I failed to find this out myself.

The legs, feet, and bill of this species are black; the lower mandible is slightly "Gull-billed" or angulated.

Total length about 15 inches.

298. STERNA CANTIACA, Gm. The Sandwich Tern.

"This Tern is seen near Tangier in abundant flocks from November to February."—*Favier*.

The Sandwich Tern is very common in the Straits in autumn, winter, and spring. Sometimes thirty or forty may be noticed sitting together on the small isolated rocks near Cabrita Point, and will allow a boat to approach within a few yards. They pass north about the first week in April, when I killed an old male tinted on the breast and under wing-coverts with a beautiful pink blush, just as is sometimes found in the spring on old males of the Black-headed Gull (*Larus ridibundus*). This Tern was also numerous at the mouth of the river at Larache during April.

This species has the bill *black, tipped with yellowish white*; legs and feet black.

Total length about 15 inches.

299. STERNA BERGII, Licht. The Swift Tern.

This large Tern was once obtained by Favier in the Straits, and is described in his MS. The specimen, which I purchased from his successor, was an adult bird in winter plumage, and is now in the possession of Lord Lilford.

The bill is pale yellow; legs and feet black.

Length 17 to 18 inches.

300. STERNA MEDIA, Horsf. The Allied Tern.

"This species is one of the least common of the Terns near Tangier, and only occasionally met with. Further south, in the vicinity of Larache, it is more frequently seen; and I found it there during September, October, and November, in company with *S. cantiaca*, which species it resembles in habits."

This Tern occurs in the Straits in spring. I obtained two, both males, shot near Tarifa on the 20th of April, 1874, and have seen others from Tangier: most probably they breed somewhere on the coast.

This bird is very much like the Sandwich Tern (*S. cantiaca*), but is a trifle larger and has the bill *yellow*. I found, on comparing male specimens shot on the same day, that it differs from that species also in having the bill stouter in proportion, and the lower mandible slightly angulated, or "Gull-billed." The feathers of the black crest are more elongated; and the upper tail-coverts and tail are grey, the same colour as the back. The primaries underneath are more broadly marked with grey next the shafts; and the tarsus is rather longer.

301. STERNA HIRUNDO, Linn. The Arctic Tern.

I obtained this Tern in one instance in winter plumage in the Straits of Gibraltar; and there is no doubt that it occurs regularly on its migration south; but it is very difficult to get specimens of the Terns in winter for identification. The Arctic Tern is distinguished from the common Tern (*S. fluviatilis*) by its much *shorter tarsus*; and the adults have the tail longer than in that species.

Total length about 15 inches, tarsus 0·55.

302. STERNA FLUVIATILIS, Naum. The Common Tern.

Favier includes this species in his notes, and states that it is found near Tangier in large flights on the coast during migration, passing south during September and October. Possibly he may mean the previous species.

The common Tern is frequently seen in autumn and spring in the Straits, and, I think, will very possibly be found nesting near Cadiz.

This species has the tarsus *long*, measuring from 0·7 to 0·75 inch; the wings reach beyond the tail.

Total length about 13 inches.

303. STERNA MINUTA, Linn. The Little Tern.

"This small Tern is seen near Tangier, passing in small flights along the coast and on the rivers and lakes. They arrive during May, and return in September, some, however, remaining in the country to breed. They all retire south for the winter."—*Favier*.

The Little Tern is only a summer visitant around Gibraltar,

and, keeping to the sea-coast, is the latest to arrive of all the family. They are nowhere very abundant; but a few nest near the mouth of the Guadiarro about the end of May, as well as in other localities on the coast.

The earliest date on which I noticed one was on the 10th of May; and the latest was on the 25th of October.

The small size of this species will distinguish it; the bill is yellow.

Total length about 8·5 inches.

304. HYDROCHELIDON HYBRIDA (Natt.). The Whiskered Tern.

"This Tern is scarce near Tangier, and seen only on passage during April, returning south in August. Immense numbers are found breeding at the lakes of Ras Dowra, where, nesting together in vast colonies, they bear the local name of *Mershik.*"—*Favier.*

On the Spanish side the Whiskered Tern arrives about the middle of April, and is seen near Gibraltar only on migration. Hovering over every swamp and wet spot, they soon pass on to their breeding-haunts in the marismas, where, among the rushes and sedges, they nest about the middle of May in colonies. Like other Terns, they lay three eggs of a pale green colour, spotted with blackish brown; but these eggs vary a great deal.

I saw a large flock of these Terns on the sea near Cadiz on the 18th of July.

In the adult birds the underparts are of a very dark grey, in some almost black.

The bill is darker in colour than the legs, which are vermilion. The total length is about 11·5 inches.

305. HYDROCHELIDON LEUCOPTERA (Meisn.). The White-winged Black Tern.

The only instance I know of the occurrence of this bird is a single specimen shot in May 1869 at Sharf el Akab, near Tangier. It is not included in Favier's list; and I never myself met with any on either side of the Straits: probably it rarely wanders so far west. I have several times imagined that I saw it, but upon shooting the birds found them to be fine old *H. hybrida*, which Tern, when flying in the bright sunshine and showing the blackish belly, is very apt to be mistaken for the present species.

The White-winged Black Tern in the adult plumage has the head, neck, and underparts all black, the vent, tail, and tail-coverts pure white.

Total length 8 to 9·5 inches.

306. HYDROCHELIDON FISSIPES (Linn.). The Black Tern.

"This Tern is abundant near Tangier when on passage, crossing the Straits in large flights during May, and returning in September and October. They are not seen in the winter months."—*Favier*.

The Black Tern, on the Spanish side, begins to appear about the end of April, and is rather later than *H. hybrida*, nesting also later in the same situations as that species, but not in such numbers. I have noticed quantities crossing the Straits on the 16th of May. They are often seen on arrival, hawking about over cornfields and low ground near water.

This Tern is more lead-coloured than black, having the tail slaty grey.

Total length about 10 inches.

Subfam. LARINÆ.

307. RISSA TRIDACTYLA (Linn.). The Kittiwake.

"This species is nearly as common during winter in the Straits as the Herring-Gull, appearing during the month of November and leaving in March."—*Favier*.

The Kittiwake is to be seen at times in great abundance during winter in the Bay of Gibraltar; at other times hardly any are to be found: their presence or absence is due to the state of the weather.

308. LARUS GELASTES, Licht. The Slender-billed Gull.

Spanish (for all Gulls). Gaviota.

Favier only records a single specimen of this Gull as obtained by him near Tangier, in 1852; but he remarks that the feet of the immature bird are orange-yellow, which would lead to the supposition that he observed it more often.

Mr. Saunders (Ibis, 1871, p. 400) states it to be by no means uncommon on the coast of Southern Spain, and suspects that it nests at the mouth of the Guadalquivir; but I can only give my own experience of never having had the good fortune to obtain it near Gibraltar.

The distinguishing mark of the species is the long slender bill; legs and feet vermilion.

Total length about 16 inches, tarsus 1·9.

309. LARUS MELANOCEPHALUS, Natt. The Mediterranean Black-headed Gull.

This Gull occasionally occurs in the Straits in winter; but I never obtained one with the black head. Mr. Saunders states, in 'The Ibis' for 1871, p. 399, that it breeds at Huelva and near Cadiz.

In the breeding-plumage the head is black.

The distinguishing mark is that the outer web of the first primary only is black.

Total length about 15 inches, tarsus 1·7.

310. LARUS RIDIBUNDUS, Linn. The Black-headed Gull.

According to Favier this Gull is the most common species around Tangier, arriving chiefly during November, and departing north in March.

The distinguishing mark of the species is that the outer web of the first quill-feather, the ends of all, and the inner webs of the other primaries are edged with black.

Total length about 15 to 16 inches, tarsus 1·6.

311. LARUS MINUTUS, Pall. The Little Gull.

Favier only mentions having once obtained this diminutive Gull near Tangier, in February 1854. I have seen it occasionally in winter; but it is not common in the Straits, though further eastward (at Malaga) Mr. Saunders states that it is not uncommon in winter and spring.

The small size of this species will serve to distinguish it, the total length being about 10·5 inches, tarsus 0·95.

This species is easily distinguished by the absence of a claw on the hind toe, which is merely a small tubercle.

Total length about 15 inches.

312. LARUS AUDOUINI, Payr.   Audouin's Gull.

This Gull is recorded by Natterer as having been once obtained near Tarifa.

Lord Lilford, who discovered it breeding in 1874 off the south of Sardinia on a rocky island, kindly gave me the following notes as to the colours of the soft parts, which marks will serve to distinguish the species.

"Feet and legs *very dark grey*, claws black. Bill brilliant *coral-red*, with *one broad black* band. Iris brilliant hazel, pupil black. Inside of mouth pale flesh-colour. Eyelids *coral-red*."

The following are the measurements (in inches) of a pair, taken from the dried skins :—

	♂	♀
Bill from gape	3·00	2·80
Wing, carpus to tip	15·75	15·75
Tarsus	2·45	2·30
Tail	6·00	6·00

The total length is about 20 inches.

313. LARUS CANUS, Linn.   The Common Gull.

This Gull is not mentioned by Favier, but is during some winters not uncommon in the Straits of Gibraltar.

Legs dark, nearly black in the adult, brownish in the young.

Total length about 18 inches; wing, carpus to tip 14·5; tarsus 1·9.

314. LARUS ARGENTATUS, Brünn.   The Herring-Gull.

The Herring Gull is stated by Favier to be "as common near Tangier during winter as *L. ridibundus*, arriving in August, September, and October, and returning north in March, April, and May."

This bird and the Lesser Black-backed Gull feed in large numbers on the refuse from the slaughterhouses at Gibraltar; and it is not uncommon to see three or four hundred of them together there.

Perhaps some of the immature birds remain during the summer; but all the adults disappear by about the 15th of April.

The legs of this species, when adult, are flesh-colour. The immature birds of this and the two next species are undistinguishable from each other.

315. LARUS LEUCOPHÆUS, Licht. Southern or Yellow-legged Herring-Gull.

Favier remarks that "this Gull is not very common near Tangier, where it consorts with *L. fuscus* and *L. argentatus.*"

Being a more eastern species, it does not often come so far as the Straits of Gibraltar; but still it is not rare.

The adult birds are distinguishable by the colour of the legs, which are yellow, as in *L. fuscus*; the eyelids are also scarlet, as in that species, the back being much lighter in colour, but darker than in *L. argentatus*.

The size of the three species is about equal.

316. LARUS FUSCUS, Linn. Lesser Black-backed Gull.

This Gull is merely included in Favier's list, but it is one of the most abundant species of Laridæ in the Straits in winter. The greater part pass northward by the end of March; but some few pairs remain to nest on the rocks of the African shore, laying about the end of April.

Total length of adult bird about 23 inches.

Legs and feet yellow; eyelids red, irides whitish yellow.

317. LARUS MARINUS, Linn. The Great Black-backed Gull.

This large Gull is, according to Favier, found about the Straits in small numbers from January to March; and he further states that he never saw any but immature birds—which agrees with my own observations.

The immature birds of this species are similar in plumage to those of *L. argentatus* and *L. fuscus*, but are distinguishable by their large size, being about 27 inches in length.

318. LARUS GLAUCUS, Gmel. The Glaucous Gull.

This northern (or, rather, Arctic) Gull is not mentioned by

Favier, but was once obtained by him in immature plumage near Tangier.

Primaries white in adult, greyish white in the immature bird.

Total length from 26½ to 33 inches.

Subfamily LESTRIDINÆ.

319. STERCORARIUS BUFFONII, Boie.   Buffon's Skua.

This Skua is recorded by Favier as twice obtained near Tangier—in 1846, and in October 1858, the first being an immature specimen. This is the smallest of the four European Skuas, and according to Mr. Howard Saunders, who kindly gave me the distinguishing marks of the Lestridinæ, has the shafts of *only* the first and second primaries white, the third dusky; the general colour of the young is also duskier.

320. STERCORARIUS PARASITICUS (Linn.).   Richardson's Skua.

Favier only mentions one specimen of this Skua, killed near Tangier in 1844. It is, however, not uncommon in winter.

This species, according to Mr. Howard Saunders, has the shafts of the first, second, third, and fourth primaries *white in all stages.*

321. STERCORARIUS POMARINUS, Temm.   Pomarine Skua.

Stated by Favier to be very rare near Tangier; and he only mentions one specimen obtained, as far back as November 1845.

I frequently saw Skuas flying about the Straits in winter, but could not succeed in shooting any. They usually appear after westerly gales, and are not seen when the weather is settled. The adult birds are very rare, and I never myself saw any.

Length about 20 inches, wing 14·25.

322. STERCORARIUS CATARRACTES (Linn.).   Common Skua.

Favier records a single specimen obtained near Tangier, in

December 1852. It occurs regularly in winter in the Straits, though not commonly.

Total length about 24 to 25 inches, wing 16.

Family PROCELLARIIDÆ.

323. PUFFINUS KUHLI, Boie.   Cinereous Shearwater.

This species of Shearwater is abundant in the Straits, and is occasionally found dead on the shore. They nest about the beginning of May, under rocks and stones on islands.

I regret to be unable to give any information about the Shearwaters, except that there appears to me to be a third species frequenting the Straits, which I could not obtain. I often went out to try and shoot these birds; but though frequently within shot when in a steamer, I never could get a chance when trying for them in a boat as is usual in such cases.

324. PUFFINUS ANGLORUM, Temm.   Manx Shearwater.

Favier states that this Shearwater is "found from August to November, and usually picked up dead on the sea-shore."

It is common in the Straits in autumn, occasionally coming close in to the land in the Bay of Gibraltar.

Head, back, wings, and tail blackish brown.

Total length about 14·5 inches, wing 9·5.

325. THALASSIDROMA PELAGICA (Linn.).   Stormy Petrel.

This little Petrel is frequently seen skimming about in the Straits, and, no doubt, nests on some of the small islands or patches of rock on the coast.

Total length 5·5 inches. Tail almost square in shape.

326. THALASSIDROMA LEUCORRHOA, Vieillot.   Fork-tailed or Leach's Petrel.

This species is stated by Favier to be of rare occurrence in the Straits. Those which he obtained were all found dead on the sea-shore after storms. He mentions picking up six in 1846, and one in each of the years 1852, 1854, and 1858.

The tail is deeply forked. Total length about 7·25 inches.

Family ALCIDÆ.

327. URIA TROILE, Temm.   The Common Guillemot.

This Guillemot is occasionally seen in small numbers about the Straits in winter, especially after severe weather from the westward.

328. ALCA TORDA, Linn.   The Razor-bill.

Moorish. Bou-drihima (*Favier*).

Favier only says of this species that it "is found near Tangier from November to February."

The Razor-bill, in some winters, appears in the Straits in very large numbers, as in the winter of 1871–72, when, during February, they were to be seen in all directions about Gibraltar Bay, some coming into the New Mole so close to the land that we threw stones at them. They lingered on very late, as I saw ten on the 19th, one on the 21st, and two on the 28th of March, and one on the 7th of April. In this case their appearance was, no doubt, attributable in the first instance to heavy gales and storms outside the Straits.

Total length about 17 inches.

329. FRATERCULA ARCTICA (Linn.).   The Puffin.

"This species is found near Tangier from November to March, sometimes even lingering as late in the spring as April and May. They are more abundant than the Gannet, and are frequently picked up dead on the sea-shore after stormy weather."—*Favier*.

I have seen Puffins in Gibraltar Bay as late as the 5th of March; but as far as my experience goes, they are a much less frequent bird than the Gannet and even than the Razor-bill (*Alca torda*).

Family COLYMBIDÆ.

330. COLYMBUS SEPTENTRIONALIS, Linn.   The Red-throated Diver.

This is the only species of Diver which I succeeded in

identifying in the Straits. I obtained one specimen with the red throat. Possibly the other two European species (*Colymbus glacialis* and *C. arcticus*) occasionally occur; but in any case some species of Diver is very common at times in winter at the entrance of the Straits.

Family PODICIPIDÆ.

331. PODICEPS CRISTATUS (Linn.). The Great Crested Grebe.

"This large Grebe is, near Tangier, less common than *Podiceps minor*. Some remain in the country to breed; the others pass north during March. They are very abundant at the lakes of Ras-Dowra."—*Favier*.

I can quite corroborate the latter statement; for when at these lakes at the end of April, the number of these Grebes, as well as of *P. griseigena*, was perfectly marvellous. They were in pairs, but had not commenced laying. These swampy lakes, much covered at the sides with aquatic plants and sedges, must be a Paradise for all Grebes and waterbirds; but it is vexation of spirit and almost useless for the ornithologist to go there. The Arabs, at the egging-season, move their tents close to the lake and plunder every nest they can find, and further pester Europeans to an unbearable degree, being almost as annoying and intrusive as the mosquitoes, which are there as troublesome as in any country I have been in. Towards Casa Vieja and Gibraltar I never met with the present species; but Lord Lilford found them breeding plentifully in May near the edge of the Coto de Doñana.

The Great Crested Grebe, in the adult breeding-plumage, is ruffed round the neck and crested; the female, smaller than the male, has the crest and frill smaller and less full.

The total length is about 21 to 22 inches.

332. PODICEPS GRISEIGENA, Bodd. The Red-necked Grebe.

"This bird is less common near Tangier than *Podiceps*

cristatus, being seldom observed on passage. Some remain in the country to breed, the others migrating northwards in March, returning again during September. They are more abundant at the lakes of Ras-Dowra, and are there called *Mazan* by the Arabs."—*Favier*.

I have seen specimens of the Red-necked Grebe obtained in Morocco by Favier so young that they must have been bred in the country; and although I was unable to procure a specimen for identification, I am confident I saw several of this species at Ras-Dowra in April. I have no record of its occurrence on the Spanish side.

The adult bird, in breeding-plumage, has no crested frill; but the neck in front is of a chestnut-red colour; this part in winter is grey.

Entire length about 16·5 inches.

333. PODICEPS AURITUS, Linn.   The Sclavonian Grebe.

Although Favier has not mentioned this Grebe as occurring near Tangier, I have seen one specimen obtained in the Straits in October 1867; and probably it is often to be met with in winter.

Slightly larger than the next species (*Podiceps nigricollis*), it is always to be distinguished, either in winter or immature plumage, by the bill, which is straight, forming an elongated cone, as in the two previous species. In the summer plumage it is frilled and crested, and may be termed a miniature Great Crested Grebe.

Total length about 13 inches.

334. PODICEPS NIGRICOLLIS, Sundev.   The Eared Grebe.

This species is the most common of the Grebes, breeding in lagoons and swamps on both sides of the Straits. In the winter they take to the salt water, and are generally plentiful in Gibraltar Bay.

The Eared Grebe is always to be distinguished by the bill, which turns slightly up, particularly the lower mandible.

In breeding-plumage the adults have a red chestnut patch

behind the eye, over the ear, the head not being frilled or crested. It is, in fact, a miniature Red-necked Grebe.

Total length about 12 inches.

335. PODICEPS MINOR, Linn. The Little Grebe or Dabchick.

Moorish. El ghotis (*Favier*).

"This small Grebe is resident near Tangier, although to a great extent migratory, passing north during April, and reappearing from October to December. It is resident and especially numerous at the lakes of Ras-Dowra, where the Arabs, during the breeding-season, in a great measure subsist on the eggs of various aquatic birds, destroying a prodigious quantity."—*Favier*.

The Dabchick is resident in Andalucia, breeding abundantly in some localities; but it is most common (or, rather, mostly noticed) in winter; and how they reach the isolated patches of water, which are dry in summer, is marvellous, as I never saw one on the wing like *Podiceps cristatus*.

The Dabchick, in winter, is almost always to be seen on the inundation at the north of Gibraltar, and takes no notice of the numerous passers by, familiarity breeding contempt.

Total length about 9·5 inches.

The following are the species included by Mr. Howard Saunders in his "List of the Birds of Southern Spain" (Ibis, 1871, pp. 54–68, 205–225, and 384–402) which have not fallen within my notice near Gibraltar; but, as will be seen, the greater part of them occurred some distance to the east of that place, far out of the district of which I have endeavoured to treat, and are here alluded to merely with the object of inviting ornithologists who may in future visit Andalucia to look out for these birds.

Doubtless many other species are to be found as stragglers; and so local are birds in Southern Spain, that perhaps some may be resident and overlooked in consequence of the exact locality they frequent having been unvisited.

1. HALIAETUS ALBICILLA, Linn. The White-tailed or Sea-Eagle.

Recorded as having been obtained near Cadiz.

Possibly this Eagle may nest on the coast where there are any cliffs suitable to its habits.

2. PICUS MEDIUS, Linn. The Middle Spotted Woodpecker.

This Woodpecker is recorded from Murcia, and, having been stated to be common in Portugal, is very probably to be met with in Andalucia.

In this species both sexes, when adult, have the crown of the head vermilion, the belly and tail-coverts rose-colour, the back being much as in *Picus major*.

Total length 8·5 inches, tarsus 0·8.

3. PICUS MINOR, Linn. The Lesser Spotted Woodpecker.

This diminutive species was first recorded from Aranjuez by Lord Lilford (*vide* Ibis, 1866, p. 183).

Total length 5·2 inches.

4. SITTA CÆSIA, Wolf. The Common Nuthatch.

This bird (our British Nuthatch) is stated by Mr. Saunders to be common near Granada. Lord Lilford mentions it as common to the north of Madrid, near La Granja.

The distinctive marks of the species are its dull brown legs and cinnamon-coloured underparts.

Total length about 5½ inches.

5. SITTA NEUMAYERI, Michah.

Sitta syriaca auctorum.

Mr. Saunders mentions having seen this bird and its nest near Archena. It is a rock-haunting Nuthatch; and hitherto the most western locality from which it has been with certainty recorded is Dalmatia. In size this species about equals *Sitta cæsia*; but it has a larger bill and feet, and lacks the white on the tail-feathers of that bird. The legs and feet also differ in being of a leaden grey colour.

6. MELIZOPHILUS SARDUS, Marm. Marmora's Warbler.

This bird is mentioned as having been only once seen, near Palma, in the island of Majorca.

It has been described as much resembling the Dartford Warbler (*Melizophilus provincialis*) in its habits, but frequenting lower ground (*cf.* Mr. A. B. Brooke, in 'The Ibis,' 1873, p. 242). That gentleman further says it is very abundant in Sardinia.

This species resembles the Dartford Warbler in size and shape, but has none of the rufous of that bird.

7. HYPOLAIS OLIVETORUM, Strickland. The Olive-tree Warbler.

This Warbler is mentioned by Mr. Saunders as seen in a museum at Valencia; but according to Mr. Dresser, in his 'Birds of Europe,' it has not been recorded from any other locality westward of its chief habitat (Greece), though it is stated to have occurred in Algeria and also in Tangier and Fez (Naumannia, ii. part i. p. 77).

The song is said to be so marked as to render it unlikely to remain unnoticed by a field-naturalist; and, like the rest of the genus, it is a tree-frequenting bird, chiefly affecting olive-groves.

The general colour of the upper parts is a dull brownish grey, the underparts being white; and it has a white eyebrow.

Total length 6 inches.

8. ACROCEPHALUS MELANOPOGON, Temm. The Moustached Sedge-Warbler.

This species is common near Valencia, and to be looked for in sedge-covered swamps and reed-beds.

It has a white eyebrow, with a dark streak below extending on both sides of the eye. The legs are said to be black.

The total length is about 5 inches.

9. ACROCEPHALUS PALUSTRIS, Crep. The Marsh-Warbler.

If this be not a race of the Reed-Warbler (*Acrocephalus streperus*), but a distinct species, it is recorded by Mr. Saunders from the neighbourhood of Madrid, at Aranjuez; but Lord Lilford mentions only the Reed-Warbler as occurring there.

10. PARUS PENDULINUS, Linn. The Penduline Titmouse.

This swamp-haunting Tit is recorded from the Albufera, near Valencia.

Length about 4·2 inches.

11. PARUS PALUSTRIS, Linn. The Marsh-Titmouse.

This well-known British bird is mentioned as found near Granada and Cordova in spring.

12. PANURUS BIARMICUS, Linn. The Bearded Titmouse.

This reed-frequenting species is recorded by Mr. Saunders as occurring near Valencia, from which locality I have seen specimens.

13. MUSCICAPA PARVA, Bechst. The Red-breasted Flycatcher.

This bird is stated to have occurred at San Roque, and to have been seen in winter near Utrera; but it does not appear by whom the specimen obtained was identified. Professor Newton remarks that the adult males of this Flycatcher have much of the appearance of the Robin (*Erythacus rubecula*); but the tail is marked with white and consists of ten feathers.

Total length about 5 inches.

14. MUSCICAPA COLLARIS, Bechst. The Collared Flycatcher.

This Flycatcher is recorded by Mr. Saunders as having been

once seen by him at Seville. If, as it has been stated to be, it is common in Portugal, it is somewhat singular that this bird should not be frequently noticed in Andalucia.

The adult male differs only from that of *M. atricapilla* in having a white band or collar on the back of the neck; the females of both species are stated to be alike.

15. LANIUS MINOR, Gm. The Lesser Grey Shrike.
Recorded from the east of Spain.
Total length about 8½ inches.

16. LANIUS NUBICUS, Licht. The Masked Shrike.
Only mentioned as once *seen* near Gibraltar.
Total length about 7 inches.

17. EMBERIZA CITRINELLA, Linn. The Yellow Bunting.
Our well-known "Yellowhammer" is stated by Mr. Saunders to occur in winter in Southern Spain.

18. EMBERIZA PYRRHULOIDES, Pall. The Marsh-Bunting.
This large race of *E. schœniclus* is recorded as met with in the east of Spain and near Valencia.

19. DRYOSPIZA CITRINELLA (Linn.). The Citril Finch.
Stated by Mr. Saunders to be common on the coast, breeding on the hills. It is, in appearance, very like the Serin Finch (*Serinus hortulanus*), but has the acute bill of the Siskin (*Chrysomitris spinus*).

20. NUCIFRAGA CARYOCATACTES (Linn.). The Nutcracker.
See *anteà*, p. 129.

21. CHETTUSIA GREGARIA (Pall.). The Black-bellied Lapwing.
Mr. Saunders mentions having seen one specimen of this bird hanging up in Cadiz market. It is an eastern species,

and was common in Oudh, having the gregarious habits of the Golden Plover (*Charadrius pluvialis*), with a very short hind toe as in the Peewit (*Vanellus cristatus*).

Total length about 12 inches.

22. ARDEA ALBA, Linn. The Great White Heron.

Mentioned as once *seen* in the marismas near Cadiz. *Vide anteà*, p. 184.

23. CICONIA ABDIMI, Licht. White-bellied Stork.

Recorded on the authority of Lopez Seoane as having been taken near Granada in June 1858.

This small Stork is mentioned by Mr. Gurney (Ibis, 1868, 257) in his notes on Mr. Layard's Birds of South Africa; and he gives a very interesting account of it. It is a very peculiarly marked species, described as having the knee-joints marked with a red or crimson band, and is figured in Rüppell's Atlas, tom. viii.

24. PELECANUS CRISPUS, Bruch. The Dalmatian Pelican.

Mr. Saunders says this bird has been obtained at Valencia and in the Balearic Islands.

One other species remains for me to notice and call attention to, viz. the Red-backed Shrike (*Lanius collurio*), which Mr. Drake included in his list of Moorish birds; but I had reasons for omitting it. However, I have since heard that it has occurred near Seville; yet till further evidence is received, I cannot place it with certainty in the List of Birds of the 'Straits.'

In elucidation of the geographical distribution of European birds, I append a *résumé* of the birds frequenting the Straits of Gibraltar. The letters indicate the precise locality of the species, as follows:—G. means the British territory of Gibraltar; M., Morocco only; S., the Spanish side of the Straits only. An asterisk (*) is affixed to all species actually shot by myself: but there are several others which I might have killed, such as the Stork, the Swallow, the Gannet, &c., but I did not care to slaughter them unnecessarily.

1. Vultur monachus.
+ 2. Gyps hispaniolensis *.
+ 3. Neophron percnopterus *. G.
4. Circus cyaneus *.
5. C. pygargus *.
6. C. macrurus.
7. C. æruginosus *.
8. Melierax polyzonus. M.
9. Astur palumbarius *.
? 10. Accipiter nisus *. G.
11. Buteo vulgaris *.
→ 12. B. desertorum *. M.
? 13. Gypaëtus barbatus.
14. Aquila chrysaëtos.
15. A. adalberti *.
16. A. rapax. M.
17. A. maculata. S.
+ 18. Nisaëtus fasciatus *. G.
19. N. pennatus *. G.
20. Circaëtus gallicus *.
+ 21. Milvus ictinus *.
+ 22. M. korschun *. G.
23. Pernis apivorus. G.
24. Elanus cæruleus.
25. Falco communis *. G.
26. F. barbarus. M.
27. F. feldeggii.
28. F. eleonoræ. M.
29. F. subbuteo.
30. F. regulus.
+ 31. Cerchneis tinnunculus *. G.
+ 32. C. naumannii *. G.
33. C. vespertinus. M.
+ 34. Pandion haliaëtus *. G.
+ 35. Strix flammea *. G.
36. Syrnium aluco.
37. Carine noctua *. G.
38. C. glaux.
39. Scops giu *. G.
40. Bubo ignavus *. G.
41. Asio otus. S.
42. A. accipitrinus.
43. A. capensis *.
44. Caprimulgus europæus.
45. C. ruficollis *. G.
46. Cypselus apus *. G.
47. C. pallidus *. G.
48. C. melba *. G.
49. Coracias garrula *.

50. Merops apiaster *. G.
51. Alcedo ispida *. G.
+ 52. Upupa epops *. G.
53. Cuculus canorus *. G.
54. Coccystes glandarius *. G.
55. Picus major *. S.
56. P. numidicus. M.
57. Gecinus sharpii *. S.
58. G. vaillantii. M.
59. G. canus. S.
60. Yunx torquilla.
+ 61. Turdus musicus *. G.
62. T. viscivorus *.
63. T. pilaris.
64. T. iliacus.
+ 65. T. merula *. G.
66. T. torquatus *.
+ 67. Petrocossyphus cyaneus *. G.
68. Monticola saxatilis. G.
69. Cinclus albicollis *.
70. Ixus barbatus *. M.
71. Crateropus fulvus. M.
72. Saxicola œnanthe *.
73. S. albicollis *. G.
74. S. stapazina *. G.
75. Dromolæa leucura *. G.
76. Pratincola rubetra *. G.
+ 77. P. rubicola *. G.
78. Philomela luscinia *. G.
+ 79. Ruticilla phœnicura *. G.
+ 80. R. tithys *. G.
81. R. moussieri. M.
82. R. wolfii *.
+ 83. Erythacus rubecula *. G.
84. Accentor collaris *. G., S.
85. A. modularis *.
86. Sylvia salicaria *. G.
+ 87. S. atricapilla *. G.
88. S. orphea *.
+ 89. S. melanocephala *. G.
90. S. curruca *. G.
91. S. rufa *. G.
92. S. conspicillata *.
93. S. subalpina *. G.
94. Melizophilus provincialis *. G.
95. Phylloscopus sibilatrix *. G.
96. P. trochilus *. G.
97. P. bonellii *. G.
+ 98. P. collybita *. G.

99. Hypolais polyglotta *.
100. H. opaca.
101. Cisticola schœnicola *.
102. Aëdon galactodes *. G.
103. Cettia sericea *.
104. Acrocephalus schœnobœnus.
105. A. aquaticus.
106. A. nævius.
107. A. luscinioides *.
108. A. streperus *.
109. A. turdoides *.
110. Regulus cristatus. S.
111. R. ignicapillus *. S.
112. Tichodroma muraria. S.
113. Certhia familiaris *.
114. Troglodytes parvulus *. G.
115. Acredula irbii *. S.
116. Parus cœruleus *. S.
117. P. teneriffæ *. M.
118. P. major *. G.
119. P. ater. S.
120. P. cristatus *. S.
121. Muscicapa atricapilla *. G.
122. Butalis grisola *. G.
123. Chelidon urbica *. G.
124. Hirundo rustica. G.
125. Cotyle riparia *. G.
126. C. rupestris *. G.
127. Lanius meridionalis *. S.
128. L. algeriensis *. M.
129. L. auriculatus *. G.
130. Telephonus erythropterus *. M.
131. Motacilla alba *. G.
132. M. yarrellii.
133. M. boarula *. G.
134. Budytes flavus *. G.
135. Anthus spinoletta *.
136. A. obscurus *.
137. A. pratensis *. G.
138. A. cervinus *.
139. A. trivialis *. G.
140. A. campestris *.
141. A. richardi *. S.
142. Alauda arvensis *.
143. Galerita machrorhyncha. M.
144. G. cristata *. G.
145. Lullula arborea.
146. Calandrella brachydactyla *. G.
147. C. bætica. S.
148. Melanocorypha calandra *. S.
149. Otocorys bilopha. M.
150. Emberiza miliaria *. G.
151. E. cirlus *.
152. E. hortulana *.
153. E. cia *. G.
154. E. schœniclus *.
155. Plectrophanes nivalis. M.
156. Fringilla cœlebs *. G., S.
157. F. spodiogena *. M.
158. F. montifringilla.
159. F. nivalis.
160. Passer montanus. S.
161. P. domesticus *. G.
162. P. salicicolus *.
163. Petronia stulta.
164. Chlorospiza chloris *. G.
165. Linota cannabina *. G.
166. L. rufescens.
167. L. montium.
168. Carduelis elegans *. G.
169. Chrysomitris spinus *. G.
170. Serinus hortulanus *. G.
171. Carpodacus githagineus. M.
172. Loxia curvirostris.
173. Coccothraustes vulgaris *.
174. Oriolus galbula *. G.
175. Corvus corax *. G.
176. C. tingitanus *. M.
177. C. frugilegus. S.
178. C. corone.
179. C. cornix. S.
180. C. monedula *.
181. Pyrrhocorax graculus *.
182. P. alpinus.
183. Pica rustica *. S.
184. P. mauritanica. M.
185. Cyanopica cookii *. S.
186. Sturnus vulgaris *.
187. S. unicolor *.
188. Garrulus glandarius *. G.
189. Columba palumbus *.
190. C. œnas *.
191. C. livia *. G.
192. Turtur auritus *.
193. T. senegalensis. M.
194. Pterocles arenarius.
195. P. alchata.
196. Caccabis petrosa *. G.
197. C. rubra *. S.
198. Coturnix vulgaris *. G.
199. Turnix sylvatica *.
200. Ortygometra crex *.
201. Porzana minuta.
202. P. marnetta *.
203. P. pygmæa *.
204. Rallus aquaticus *.
205. Gallinula chloropus *.
206. Fulica atra *.
207. F. cristata.
208. Porphyrio hyacinthinus *.
209. Otis tarda *.
210. Tetrax campestris *.

211. Eupodotis arabs.   M.
212. Houbara undulata.   M.
213. Œdicnemus crepitans *.
214. Glareola pratincola *.
215. Cursorius gallicus.   M.
+ 216. Vanellus cristatus *.
217. Squatarola helvetica *.   G.
+ 218. Charadrius pluvialis *.   G.
219. Eudromias morinellus.
+ 220. Ægialitis hiaticula *.   G.
+ 221. Æ. fluviatilis *.   G.
222. Æ. cantiana *.   G.
223. Strepsilas interpres *.   G.
+ 224. Hæmatopus ostralegus.
225. Recurvirostra avocetta.
226. Himantopus candidus *.
227. Phalaropus fulicarius.
228. Totanus canescens *.
229. T. fuscus.
230. T. calidris *.
231. T. glareola *.
? 232. T. ochropus *.
+ 233. T. hypoleucus *.   G.
234. Limosa ægocephala *.
235. L. lapponica *.
236. Machetes pugnax *.
237. Tringa canutus.
238. T. nigricans.   G.
239. T. subarquata *.   G.
240. T. minuta *.
241. T. temmincki *.
242. T. cinclus *.   G.
243. Calidris arenaria *.   G
244. Gallinago gallinula *.
245. G. media *.
246. G. major *.
247. Scolopax rusticola *.   G.
248. Numenius arquata.
249. N. phæopus.
250. N. hudsonicus.
251. N. tenuirostris.
252. Grus communis *.
253. G. virgo.
254. Ardea purpurea *.
255. A. cinerea *.
256. Herodias garzetta *.
257. Ardeola russata *.
258. A. comata *.
259. Ardetta minuta *.
260. Nycticorax griseus *.
261. Botaurus stellaris *.
+ 262. Ciconia alba.
263. C. nigra.
264. Ibis falcinellus.
265. Platalea leucorodia *.
+ 266. Phœnicopterus antiquorum *.

267. Cygnus musicus.
268. C. olor.   M.
269. Anser cinereus *.
270. A. segetum.
271. Bernicla leucopsis.   S.
272. Tadorna vulpauser.
273. T. rutila *.
274. Spatula clypeata *.
+ 275. Chaulelasmus streperus *.
+ 276. Anas boschas *.
277. A. angustirostris *.
278. Querquedula circia *.
+ 279. Q. crecca *.
280. Dafila acuta *.
281. Mareca penelope *.
282. Nyroca ferruginea *.
283. Fuligula rufina.
284. F. ferina *.
285. F. marila.
286. F. cristata.
287. Glaucion clangula.
288. Erismatura mersa.
289. Œdemia nigra.
290. Mergus albellus.
291. M. serrator.
292. M. merganser.
293. Phalacrocorax carbo.
+ 294. P. graculus *.
+ 295. Sula bassana.
296. Sterna caspia.
297. S. anglica *.
298. S. cantiana *.
299. S. bergii.
300. S. media.
301. S. hirundo.
302. S. fluviatilis.
303. S. minuta.
304. Hydrochelidon hybrida *.
305. H. leucoptera.
306. H. fissipes *.
+ 307. Rissa tridactyla.
308. Larus gelastes.
309. L. melanocephalus.
+ 310. L. ridibundus *.
311. L. minutus.
312. Larus audouini.
313. L. canus.
+ 314. L. argentatus *.
315. L. leucophœus *.
+ 316. L. fuscus °.
+ 317. L. marinus.
318. L. glaucus.
319. Stercorarius buffoni.
320. S. parasiticus.
321. S. pomarinus.
322. S. catarrhactes.

323. Puffinus kuhli.
324. P. anglorum.
325. Thalassidroma pelagica.
326. T. leucorrhoa.
327. Uria troile.
328. Alca torda.
329. Fratercula arctica.

330. Colymbus septentrionalis.
331. Podiceps cristatus.
332. P. griseigena. M.
333. P. auritus.
334. P. nigricollis.
335. P. minor *. G.

INDEX.

Accentor, Alpine, 84.
——, Hedge-, 84.
Accentor collaris, 84.
—— modularis, 84.
Accipiter nisus, 36.
Acredula irbii, 99.
—— vagans, 99.
Acrocephalus aquaticus, 94.
—— arundinaceus, 97.
—— luscinioides, 81, 94.
—— melanopogon, 223.
—— nævius, 94.
—— palustris, 224.
—— schœnobænus, 94.
—— streperus, 97, 224.
Actiturus bartrami, 171.
Aedon galactodes, 92.
Ægialitis cantiana, 162, 174.
—— fluviatilis, 162.
—— hiaticula, 161, 162, 163, 173.
Alauda arvensis, 112.
Alca torda, 218.
Alcedo ispida, 66.
Anas angustirostris, 199.
—— boschas, 199, 201.
Andalucian Quail, 13.
Anser albifrons, 195, 196.
—— brachyrhynchus, 195, 196.
—— cinereus, 20, 194, 196.
—— segetum, 20, 194, 195, 196.
Anthus campestris, 111.
—— cervinus, 110.
—— obscurus, 110.
—— pratensis, 110.
—— richardi, 111.
—— spinoletta, 110.
—— trivialis, 111.

apus, Cypselus, 63.
Aquila adalberti, 29, 39, 40, 45.
—— chrysaëtos, 39.
—— imperialis, 39, 40.
—— maculata, 40.
—— rapax, 40.
Ardea alba, 226.
—— cinerea, 183.
—— purpurea, 182, 186.
Ardeola comata, 186.
—— russata, 108, 185.
Ardetta minuta, 187.
Asio accipitrinus, 60.
—— capensis, 61.
—— otus, 60.
Astur palumbarius, 36.
Athene glaux, 58.
Avocet, 164.

Bee-eater, 5, 13, 15, 65.
Bernicla leucopsis, 196.
Bittern, Common, 188.
——, Little, 187.
Blackbird, 74.
Blackcap, 84, 86.
Botaurus stellaris, 187, 188.
Brambling, 118.
Bubo ignavus, 59.
Budytes flavus vel cinerescapillus, 109.
—— rayi, 110.
Bulbul, Dusky, 76.
Bullfinch, Desert, 123.
Bunting, Cirl, 116.
——, Common, 115.
——, Marsh-, 225.
——, Ortolan, 116.
——, Reed-, 117.
——, Rock-, 116.
——, Snow-, 117.

Bunting, Yellow, 225.
Bustard, Great, 22, 147, 148.
——, Houbara, 153.
——, Little, 16, 22, 34, 150.
——, North-African, 152.
——, Ruffed, 153.
Bustards, 22.
Butcher-bird, 107.
Buteo desertorum, 37.
—— vulgaris, 31, 37, 51.
Buzzard, Common, 37, 38.
——, Honey, 14, 49.
——, Rufous, 37.

Caccabis petrosa, 136, 138.
—— rubra, 21, 137.
Calandrella bætica, 114.
—— brachydactyla, 110, 114.
Calidris arenaria, 174.
Caprimulgus europæus, 62.
—— ruficollis, 62.
Carduelis elegans, 24, 122.
Carine glaux, 58.
—— noctua, 58, 59.
Carpodacus erythrinus, 124.
—— githagineus, 123.
Cerchneis naumanni, 51.
—— tinnunculus, 53.
—— vespertina, 54.
Certhia familiaris, 98, 101.
Cettia sericea, 93.
Chaffinch, Common, 118.
——, North-African, 118.
Charadrius fulvus, 161.

Charadrius pluvialis, 160, 226.
Chaulelasmus streperus, 198.
Chelidon urbica, 104, 105.
Chettusia gregaria, 225.
Chiff-chaff, 89.
Chlorospiza chloris, 120, 124.
—— chlorotica, 121.
Chough, Alpine, 128.
——, Common, 128.
Choughs, 9.
Chrysomitris spinus, 123.
Ciconia abdimi, 226.
—— alba, 188, 190.
—— nigra, 190.
Cinclus albicollis, 76.
Circaëtus gallicus, 31, 46.
Circus æruginosus, 33.
—— cineraceus, 32.
—— cyaneus, 32, 33.
—— macrurus, 32, 33.
—— pygargus, 32, 33.
Cisticola schœnicola, 91.
Coccothraustes vulgaris, 124.
Coccystes glandarius, 69.
Columba livia, 11, 51, 133.
—— œnas, 133.
—— palumbus, 132.
Colymbus septentrionalis, 218.
Coot, Bald, 145.
——, Common, 145, 146.
——, Crested, 145.
——, Red-lobed, 145.
Coracias garrula, 65.
Cormorant, Common, 207.
——, Green, 207.
Corn-Crake, 142.
Corvus corax, 126.
—— cornix, 128.
—— corone, 127.
—— frugilegus, 127.
—— monedula, 129.
—— tingitanus, 126.
Coturnix vulgaris, 138, 139.
Cotyle riparia, 104.
—— rupestris, 103, 104.
Courser, Cream-coloured, 155.
Crake, Baillon's, 142, 143.
——, Little, 142.
——, Spotted, 143.
Crane, 14, 169.
——, Common, 14, 179.
——, Demoiselle, 181.
——, Numidian, 181.

Crateropus fulvus. 78.
Creeper, Tree-, 13, 98.
——, Wall-, 98.
Crossbill, Common, 124.
Crow, Carrion-, 127.
——, Hooded, 128.
Cuckoo, Common, 69.
——, Great Spotted, 69, 129.
Cuckoos, 15.
Cuculus canorus, 69.
Curlew, Common, 178.
——, Slender-billed, 179.
——, Stone-, 153.
Curlews, 21.
Curruca atricapilla, 85.
Cursorius gallicus, 155.
—— isabellinus, 161.
Cyanopica cookii, 70, 129.
Cygnus musicus, 194.
—— olor, 194.
Cypselus melba, 64.
—— pallidus, 63.

Dabchick, 221.
Dafila acuta, 201.
Dipper, 76.
Diver, Red-throated, 218.
Dotterel, 161.
——, Ringed, 161, 163, 173.
Dove, Ring-, 132.
——, Rock-, 11, 133.
——, Stock-, 133.
Dromolæa leucura, 76, 79.
Dryospiza citrinella, 123, 225.
Duck, Brahminy, 197.
——, Ferruginous, 204.
——, Marbled, 199.
——, Scaup, 203.
——, Tufted, 203.
——, White-headed, 205.
——, Wild, 199, 201.
Dunlin, 172, 173.

Eagle, Bonelli's, 40, 46, 51.
——, Booted, 45.
——, Golden, 9. 39.
——, Imperial, 13.
——, Sea-, 222.
——, Short-toed, 46.
——, Spotted, 40.
——, Tawny, 40.
——, White-shouldered, 39.
——, White-tailed, 222.
Egret, Little, 184.

Elanus cæruleus, 50.
Emberiza cia, 116.
—— cirlus, 116.
—— citrinella, 116, 225.
—— hortulana, 116.
—— miliaria, 115.
—— pyrrhuloides, 225.
—— schœniclus, 117.
Erismatura leucocephala, 205.
Erythacus rubecula, 83, 224.
Eudromias morinellus, 161.
Eupodotis arabs, 152.
—— cristata, 153.
—— edwardsii, 152.

Falco æsalon, 52.
—— barbarus, 51.
—— communis, 50.
—— eleonoræ, 52.
—— feldeggi, 51.
—— subbuteo, 52.
Falcon, Barbary, 51,
——, Eleonora. 52.
——, Peregrine, 50.
Fieldfare, 73.
Finch, Citril, 123, 225.
——, Mountain-, 118.
——, Serin, 123.
——, Snow-, 119.
Finches, 22.
Flamingo, 193.
Flycatcher, Collared, 102, 224.
——, Pied, 101.
——, Red-breasted, 224.
Francolinus bicalcaratus, 138.
Fratercula arctica, 218.
Fringilla cœlebs, 118.
—— incerta, 124.
—— montifringilla, 118.
—— nivalis, 119.
—— spodiogena, 118.
Fulica atra, 145.
—— cristata, 145, 203.
—— ferina, 203, 204.
—— rufina, 198, 202, 204.
—— marila, 203.

Gadwall, 198.
Galerita cristata, 112.
—— macrorhyncha, 112.
Gallinago gallinula, 174.
—— major, 176.
—— media, 175.
Gallinula chloropus, 144.
Gannet, 55, 207.

INDEX. 233

Garrulus glandarius, 132.
Gecinus canus, 72.
—— sharpii, 5, 71, 72.
—— vaillantii, 72.
—— viridis, 71, 72.
Geese, Bean-, 20.
——, Wild, 20.
Glareola pratincola, 154.
—— torquata, 141.
Glaucion clangula, 205.
Goat-sucker, Common, 62.
——, Rufous-naped, 62.
Godwit, 5.
——, Bar-tailed, 170.
——, Black-tailed, 169, 170.
Golden-eye, 205.
Goldfinch, 24, 122.
Goosander, 206.
Goose, Bean-, 195, 196.
——, Bernicle, 196.
——, Grey Lag, 20, 194, 196.
——, Pink-footed, 195, 196.
——, Solan, 207.
——, White-fronted, 195, 196.
Goshawk, 36.
Grebe, Eared, 220.
——, Great Crested, 219, 220.
——, Little, 221.
——, Red-necked, 219.
——, Sclavonian, 220.
Greenfinch, 120, 124.
Greenshank, 165.
Grosbeak, Scarlet, 124.
Grouse, Black-bellied Sand-, 135.
——, Pin-tailed Sand-, 135.
Grus communis, 169, 179.
—— virgo, 181.
Guillemot, Common, 218.
Gull, Audouin's, 214.
——, Black-headed, 209, 213.
——, Common, 214.
——, Glaucous, 215.
——, Great Black-backed, 215.
——, Herring-, 212, 214.
——, Lesser Black-backed, 215.
——, Little, 213.
——, Mediterranean Black-headed, 213.

Gull, Slender-billed, 212.
——, Southern, 215.
——, Yellow-legged Herring-, 215.
Gulls, 13.
Gypaëtus barbatus, 38.
Gyps fulvus, 29.
—— hispaniolensis, 29.

Hæmatopus ostralegus, 163.
Haliaëtus albicilla, 222.
Harrier, Hen-, 32.
——, Marsh-, 19, 32, 33, 34.
——, Montagu's, 25.
——, Pale-chested, 33.
Hawfinch, Common, 124.
Hawk, Cuckoo-, 36.
——, Many-banded, 35.
Herodias alba, 184.
—— garzetta, 184, 186.
Heron, Buff-backed, 185.
——, Common, 183.
——, Great White, 184, 226.
——, Night-, 13, 187.
——, Purple, 182.
——, Squacco, 186.
Himantopus candidus, 164.
Hirundo rustica, 102, 103.
Hobby, 52, 54.
Hooper, 194.
Hoopoe, 5, 13, 68.
Houbara undulata, 153.
Hydrochelidon fissipes, 212.
—— hybrida, 211, 212.
—— leucoptera, 211.
Hypolais elaica, 91.
—— icterina, 90.
—— olivetorum, 223.
—— polyglotta, 90, 91.

Ibis, Glossy, 191.
Ibis falcinellus, 191.
Ixus barbatus, 76.

Jackdaw, 128.
Jacksnipe, 174.
Jay, Common, 132.

Kestrel, Common, 53.
——, Lesser, 51, 53.
Kingfisher, 68.
Kite, Common, 38.

Kite, Black, 14, 38, 48.
——, Black-shouldered, 50.
——, Marbled, 36.
Kites, 14.
Kittiwake, 212.
Knot, 171.

Lammergeyer, 5, 38.
Landrail, 142.
Lanius algeriensis, 106, 107.
—— auriculatus, 106.
—— collurio, 107, 226.
—— meridionalis, 105, 106, 107.
—— minor, 225.
—— nubicus, 225.
Lanner, 51.
Lapwing, 158.
——, Black-bellied, 225.
Lark, Andalucian Short-toed, 114.
——, Calandra, 114.
——, Crested, 112.
——, Horned Desert-, 115.
——, Short-toed, 114.
——, Sky-, 112.
——, Tristram's, 112.
——, Wood-, 113.
Larks, 22.
Larus argentatus, 214, 215.
—— audouini, 214.
—— canus, 214.
—— fuscus, 215.
—— gelastes, 212.
—— glaucus, 215.
—— leucophæus, 215.
—— marinus, 215.
—— minutus, 213.
—— melanocephalus, 213.
—— ridibundus, 209, 213, 214.
Limosa ægocephala, 169, 170.
—— lapponica, 170.
Linnet, Common, 121.
——, Mountain-, 121.
Linota rufescens, 121.
—— montium, 121.
Loxia curvirostra, 124.
Lullula arborea, 113.
Lymnocryptes, 175.

Machetes pugnax, 170.
Magpie, Azure-winged, 129.
——, Common, 129.

Magpie, North-African, 129.
——, Spanish, 129.
Mareca penelope, 202.
Martin, Crag-, 104.
——, House-, 102.
——, Rock, 104.
——, Sand-, 104.
Melanocorypha calandra, 114.
Melierax polyzonus, 35.
Melizophilus provincialis, 223.
—— sardus, 223.
—— undatus, 88.
Merganser, Red-breasted, 206.
Mergus albellus, 206.
—— merganser, 206.
—— serrator, 206.
Merlin, 52.
Merops apiaster, 65.
Milvus ictinus, 31, 38, 47.
—— korschun, 14, 31, 37, 38, 48.
Monticola saxatilis, 76.
Moorhen, 144.
Motacilla alba, 108.
—— sulfurea, 108.
—— yarrellii, 108.
Muscicapa albicollis, 102.
—— atricapilla, 101, 225.
—— collaris, 224.
—— parva, 224.

Neophron, 5, 13, 14, 29, 31, 39.
—— percnopterus, 31.
Nightingale, 80.
Nightingales, 13.
Nightjar, 62.
Nisaëtus fasciatus, 40.
—— pennatus, 31, 45, 47.
Nucifraga caryocatactes, 225, 129.
Numenius arquata, 178.
—— hudsonicus, 178.
—— phæopus, 178.
—— tenuirostris, 179.
Nutcracker, 129, 225.
Nuthatch, Common, 222.
Nycticorax griseus, 187.
Nyroca ferruginea, 204.

Œdemia fusca, 206.
—— nigra, 205.
Œdicnemus crepitans, 153, 164.
Oriole, Golden, 125.
Oriolus galbula, 125.
Ortygometra crax, 142.

Osprey, 44, 45, 51, 54.
Otis tarda, 147.
Otocorys bilopha, 115.
Otogyps auricularis, 31.
Ouzel, Ring-, 74.
——, Water-, 76.
Owl, Barn-, 56.
——, Cape-, 61.
——, Eagle, 13, 59.
——, Short-eared, 60.
——, Long-eared, 60.
——, Marsh-, 61.
——, Scops, 58.
——, Southern Little, 58.
——, Tawny, 57.
——, White, 56.
Owls, 4, 12.
Oystercatcher, 163.

Pandion haliaëtus, 54.
Panurus biarmicus, 224.
Partridge, Barbary, 13, 15, 136.
——, Common Red-legged, 13, 137.
——, French, 137.
Parus ater, 101.
—— cæruleus, 100.
—— ledouci, 101.
—— major, 5, 100.
—— palustris, 101, 224.
—— pendulinus, 224.
—— teneriffæ, 100.
Passer cisalpinus, 119.
—— domesticus, 119.
—— montanus, 119.
—— salicicolus, 120.
Peewit, 13, 21, 158, 169, 226.
Pelecanus crispus, 226.
Pelican, Dalmatian, 226.
Pernis apivorus, 49.
Petrel, Fork-tailed, 217.
——, Leach's, 217.
——, Stormy, 217.
Petrocossyphus cyanus, 74.
Petronia stulta, 120.
Petrosa rubra, 15.
Phalacrocorax carbo, 207.
Phalarope, Grey, 165.
Phalaropus fulicarius, 165.
Philomela luscinia, 80.
Phœnicopterus antiquorum, 193.
Phylloscopus Bonellii, 89.
—— collybita, 89.
—— rufus, 89.
—— sibilatrix, 88.
—— trochilus, 89.

Pica rustica, 70, 129.
Picus major, 5, 70, 71, 222.
—— mauritanica, 129.
—— medius, 222.
—— minor, 222.
—— numidicus, 71.
Pintail, 201.
Pipit, Meadow-, 110.
——, Red-throated, 110.
——, Richard's, 111.
——, Rock-, 110.
——, Tawny, 111.
——, Tree-, 111.
——, Water-, 110.
Platalea leucorodia, 192.
Plectrophanes nivalis, 117.
Plover, Eastern Golden, 161.
——, Golden, 13, 21, 131, 160, 226.
——, Grey, 159.
——, Kentish, 162, 174.
——, Little Ringed, 162.
——, Ringed, 161.
Pochard, Common, 203.
——, Red-crested, 198, 202.
——, Red-headed, 203.
——, White-eyed, 204.
Podiceps auritus, 220.
——, cristatus, 219, 221.
——, griscigena, 219.
——, minor, 219, 220.
——, nigricollis, 220.
Porphyrio hyacinthinus, 146.
Porzana maruetta, 143, 144.
—— minuta, 142.
—— pygmæa, 143.
Pratincola rubetra, 80.
Pratincole, Collared, 154.
Pterocles alchata, 135.
—— arenarius, 135.
—— senegallus, 136.
Puffin, 218.
Puffinus anglorum, 217.
——, kuhli, 217.
Pyrrhocorax alpinus, 128.
—— graculus, 128, 207.

Quail, 14, 16, 21.
——, Andalucian Bush-, 139.
——, Bush-, 141.
——, Button-, 141.
——, Common, 138.
——, Indian Bush-, 140.
——, Three-toed, 141.

INDEX.

Querquedula circia, 200.
—— crecca, 201.

Rail, Water-, 144.
Rallus aquaticus, 144.
Raven, Common, 126.
——, Tangier, 126.
Ravens, 44.
Razorbill, 218.
Recurvirostra avocetta, 164.
Redpole, Lesser, 121.
Redshank, Common, 166.
——, Dusky, 166.
——, Spotted, 166.
Redstart, Black, 82.
——, Common, 81.
——, Moussier's, 82.
Redwing, 74.
Reeve, 170.
Regulus cristatus, 98.
——, ignicapillus, 98.
Rissa tridactyla, 212.
Robin, 83.
Roller, 65.
Rook, 127.
Ruff, 170.
Ruticilla erythrogastra, 82.
—— moussieri, 82.
—— phœnicura, 81.
—— titys, 81, 82.
—— wolfii, 82.

Sanderling, 174.
Sandgrouse, 135.
Sandpiper, Bartram's, 171.
——, Common, 168.
——, Curlew, 172.
——, Green, 5, 167.
——, Marsh-, 166.
——, Purple, 168, 171.
——, Wood-, 167.
——, Pygmy, 172.
Saxicola albicollis, 79.
—— œnanthe, 78.
—— stapazina, 79.
Scolopax sabinii, 176.
—— rusticola, 177.
Scops giu, 58.
Scoter, Common, 205.
——, Velvet, 206.
Serinus hortulanus, 123.
Shag, 207.
Shearwater, Cinereous, 217.
——, Manx, 217.
Sheldrake, Common, 196.
——, Ruddy, 197.
Shrike, Algerian Grey, 106.

Shrike, Hooded, 107.
——, Lesser Grey, 225.
——, Masked, 225.
——, Red-backed, 226.
——, Spanish Grey, 105.
——, Woodchat, 106.
Shoveller, 197.
Siskin, 123.
Sitta cæsia, 222.
—— neumayeri, 222.
—— syriaca, 222.
Skua, Buffon's, 216.
——, Common, 216.
——, Pomarine, 216.
——, Richardson's, 216.
Smew, 206.
Snipe, 15, 17, 18, 19.
——, Common, 175, 177.
——, Great, 176.
——, Jack, 18.
——, Solitary, 176.
Sparrow, Common, 119.
——, Rock-, 120.
——, Spanish, 120.
——, Tree-, 119.
Sparrowhawk, 36.
Spatula clypeata, 197.
Spoonbill, 191.
——, Common, 192.
——, White, 192.
Squatarola helvetica, 159.
Starling, Common, 130.
——, Sardinian, 131.
Stercorarius Buffonii, 216.
—— catarractes, 216.
—— parasiticus, 216.
—— pomarinus, 216.
Sterna anglica, 208.
—— bergii, 209.
—— cantiaca, 209.
—— caspia, 208.
—— fluviatilis, 210.
—— hirundo, 210.
—— media, 209.
—— minuta, 210.
Stilt, Black-winged, 164.
Stint, Little, 172, 173.
——, Temminck's, 173.
Stonechat, 80.
Stone-Curlew, 5.
Stork, Black, 190.
——, White, 77, 188, 190.
——, White-bellied, 226.
Strepsilas interpres, 163.
Strix flammea, 56.
—— noctua meridionalis, 58.
Sturnus unicolor, 131.
——, vulgaris, 130.
Sula bassana, 207.

Swallow, Common, 103.
Swallows, 13.
Swan, Mute, 194.
——, Whistling, 194.
Swift, Alpine, 64.
——, Common, 63.
——, Mouse-coloured, 63.
——, White-bellied, 64.
Sylvia atricapilla, 84.
—— conspicillata, 87.
—— curruca, 86.
—— melanocephala, 84, 85.
—— orphea, 85.
—— rufa, 87.
—— salicaria, 84.
—— subalpina, 87.
Syrnium aluco, 57.

Tadorna rutila, 197.
—— vulpanser, 196.
Teal, Common, 201.
——, Garganey, 200.
Telephonus erythropterus, 107.
Tern, Allied, 209.
——, Arctic, 210.
——, Black, 212.
——, Caspian, 208.
——, Common, 210.
——, Gull-billed, 208.
——, Little, 210.
——, Sandwich, 209.
——, Swift, 209.
——, White-winged Black, 211.
Terns, 13.
Tetrax campestris, 150.
Thalassidroma leucorrhoa, 217.
—— pelagica, 217.
Thick-knee, 153.
Thrush, Algerian Babbling, 78.
——, Blue, 13.
——, Blue Rock-, 74.
——, Common, 73.
——, Missel-, 73.
——, Rock-, 76.
Tichodroma muraria, 98.
Titmouse, Algerian Cole, 101.
——, Bearded, 224.
——, Blue, 100.
——, European Cole, 101.
——, Great, 100.
——, Marsh-, 224.
——, Penduline, 224.
——, Spanish Long-tailed, 99.

INDEX.

Titmouse, Ultramarine, 100.
Totanus calidris, 166.
—— canescens, 165.
—— fuscus, 166.
—— glareola, 167.
—— hypoleucus, 168.
—— ochropus, 167.
—— stagnatilis, 166, 167.
Tringa canutus, 160, 171.
—— cinclus, 172, 173.
—— maritima, 168.
—— minuta, 172, 173.
—— nigricans, 171.
—— schinzii, 173.
—— subarquata, 160, 161.
—— temminckii, 173.
Troglodytes parvulus, 99.
Turdus iliacus, 74.
—— merula, 74.
—— musicus, 73.
—— pilaris, 73.
—— torquatus, 74.
—— viscivorus, 73.
Turnix sylvatica, 139.
Turnstone, 163.
Turtledove, Common, 134.
——, Egyptian, 135.
Turtur auritus, 134.
—— senegalensis, 135.
Twite, 121.

Upupa epops, 68.
Uria troile, 218.

Vanellus cristatus, 158, 169, 226.
Vultur monachus, 28.

Vulture, Black, 28.
——, Egyptian, 31.
——, Griffon, 13, 37.
——, Sociable, 31.

Wagtail, 5.
——, Grey, 108.
——, Grey-headed Yellow, 109.
——, Pied, 108.
——, White, 108.
Warbler, Aquatic, 94.
——, Black-headed, 13, 85.
——, Blue-throated, 82.
——, Bonelli's Willow-, 80.
——, Cetti's, 93.
——, Dartford, 88, 223.
——, Fantail, 91.
——, Garden-, 84, 86.
——, Grasshopper, 94.
——, Great Sedge-, 97.
——, Marmora's, 223.
——, Melodious Willow-, 90.
——, Marsh-, 224.
——, Moustached Sedge-, 223.
——, Olive-tree, 223.
——, Orphean, 85.
——, Reed-, 97, 224.
——, Rufous, 92.
——, Savi's, 87, 94, 115.
——, Sedge-, 94.
——, Spectacled, 87.
——, Subalpine, 87.
——, Western Pallid, 91.

Warbler, Willow-, 89.
——, Wood-, 88.
——, Yellow Willow-, 90.
Waterhen, 144.
——, Purple, 146.
Wheatear, 4, 13.
——, Black, 79.
——, Common, 78.
——, Eared, 79.
——, Russet, 79.
Whimbrel, 178.
——, American, 178.
Whinchat, 80.
Whitethroat, Common, 87.
——, Lesser, 86.
Widgeon, 202.
Woodcock, 16, 177.
Woodpecker, Algerian Green, 72.
——, Algerian Pied, 71.
——, Great Spotted, 70.
——, Grey-headed Green, 72.
——, Lesser Spotted, 222.
——, Middle Spotted, 222.
——, Spanish Green, 71.
Woodpeckers, 13, 21.
Woodpigeon, 132.
Wren, 99.
——, Fire-crested, 98.
——, Gold-crested, 98.
Wryneck, 72.

Yunx torquilla, 72.

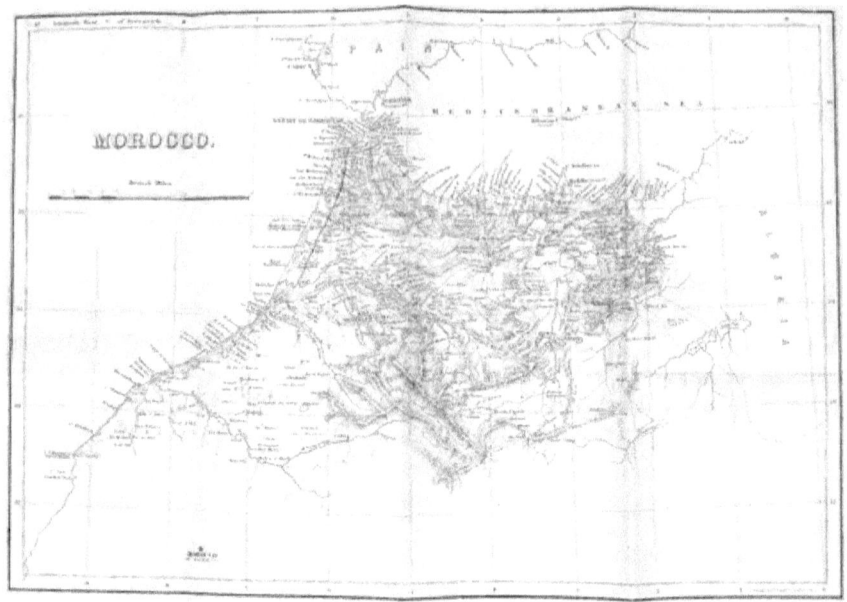

www.ingramcontent.com/pod-product-compliance
Lightning Source LLC
Chambersburg PA
CBHW031736230426
43669CB00007B/359